Controlling Pollution in Transition Economies

NEW HORIZONS IN ENVIRONMENTAL ECONOMICS

General Editor: Wallace E. Oates, *Professor of Economics, University of Maryland*

This important series is designed to make a significant contribution to the development of the principles and practices of environmental economics. It includes both theoretical and empirical work. International in scope, it addresses issues of current and future concern in both East and West and in developed and developing countries.

The main purpose of the series is to create a forum for the publication of high quality work and to show how economic analysis can make a contribution to understanding and resolving the environmental problems confronting the world in the late twentieth century.

Recent titles in the series include:

The International Yearbook of Environmental and Resource Economics
1997/1998
A Survey of Current Issues
Edited by Henk Folmer and Tom Tietenberg

The Economic Theory of Environmental Policy in a Federal System
Edited by John B. Braden and Stef Proost

Environmental Taxes and Economic Welfare
Reducing Carbon Dioxide Emissions
Antonia Cornwell and John Creedy

Economics of Ecological Resources
Selected Essays
Charles Perrings

Economics for Environmental Policy in Transition Economies
An Analysis of the Hungarian Experience
Edited by Péter Kaderják and John Powell

Controlling Pollution in Transition Economies
Theories and Methods
Edited by Randall Bluffstone and Bruce A. Larson

Environments and Technology in the Former USSR
Malcolm R. Hill

Pollution and the Firm
Robert E. Kohn

Climate Change, Transport and Environmental Policy
Empirical Applications in a Federal System
Edited by Stef Proost and John B. Braden

The Economics of Energy Policy in China
Implications for Global Climate Change
ZhongXiang Zhang

Advanced Principles in Environmental Policy
Anastasios Xepapadeas

Taxing Automobile Emissions for Pollution Control
Maureen Sevigny

Controlling Pollution in Transition Economies

Theories and Methods

Edited by

Randall Bluffstone

Project Associate, Harvard Institute for International Development, Harvard University, USA

and

Bruce A. Larson

Project Associate, Harvard Institute for International Development, Harvard University, USA

NEW HORIZONS IN ENVIRONMENTAL ECONOMICS

Edward Elgar
Cheltenham, UK • Lyme, US

Published by
Edward Elgar Publishing Limited
8 Lansdown Place
Cheltenham
Glos GL50 2HU
UK

Edward Elgar Publishing, Inc.
1 Pinnacle Hill Road
Lyme
NH 03768
US

A catalogue record for this book
is available from the British Library

Library of Congress Cataloguing in Publication Data

Controlling pollution in transition economies : theories and methods /
    edited by Randall Bluffstone and Bruce A. Larson.
        (New horizons in environmental economics)
    Includes index.
        1. Environmental impact charges—Europe, Eastern.
    2. Environmental impact charges—Former Soviet republics.
    3. Pollution—Government policy—Europe, Eastern.   4. Pollution—
    Government policy—Former Soviet republics.   I. Bluffstone,
    Randall, 1960–   .   II. Larson, Bruce A.   III. Series.
    HJ5403.57.Z73C66   1997
    363.73'7'0947—dc21   97–25020
                                                                    CIP

Printed and bound in Great Britain by
Hartnolls Limited, Bodmin, Cornwall

ISBN 1 85898 452 1

Contents

Figures

Tables

Contributors

Editors

Randall Bluffstone, Project Associate, Harvard Institute for International Development. Dr. Bluffstone has been with the International Environment Program of the Harvard Institute for International Development since January, 1993, and during the period 1994–1997 served as Senior Environmental Policy Advisor to the government of Lithuania. His research has focused on economic instruments for environmental management, privatization and the environment, environmental finance, and deforestation in low-income countries.

Bruce A. Larson, Project Associate, Harvard Institute for International Development. Dr. Larson has had long-term assignments implementing environmental economic and policy analysis programs in Estonia, Latvia and Russia. Prior to joining the Harvard Institute for International Development in 1994, Dr. Larson was an economist with the United States Department of Agriculture. He has also been a visiting fellow with the Winrock Institute for Agricultural Development.

Non-Country Contributors

Jeffrey R. Vincent, Fellow of the Institute, Harvard Institute for International Development. Since 1994, Dr. Vincent has been the director of the Harvard Institute for International Development's Environmental Economics and Policy Project in the Newly Independent States. In the transition economies, Dr. Vincent has research and advisory experience in Bulgaria, Kazakhstan, Romania and Russia. He is also the author of a forthcoming book to be published by Harvard University Press titled *Environment and Development in a Resource-Rich Economy: Malaysia Under the New Economic Policy.*

Scott Farrow, Senior Economist, Dames and Moore, Inc. and the Heinz School of Public Policy and Management, Carnegie Mellon University.

Dr. Farrow has conducted research and consulting projects on the following topics: benefit-cost analysis for project and regulatory review, optimal extraction, the use of economic instruments for environmental management, and marine energy resources.

Bulgaria

Nino I. Ninov, Engineer, Department of Air Protection, Ministry of Environment of Bulgaria.

Nikola Matev, Senior Specialist, Department of Water Protection, Ministry of Environment of Bulgaria.

Czech Republic

Zdenek Stepanek, Environmental Economist and Executive Director of the consulting company TECHEM. Prior to founding TECHEM, Dr. Stepanek was Director of the Economic Policy Department of the Ministry of Environment of the Czech Republic.

Estonia

Ljuba Gornaja, Environmental Economics Advisor in the Environmental Information Centre of the Ministry of Environment of Estonia. For the last eight years Dr. Gornaja has contributed to the development and implementation of the Estonian system of permits and charges. She was previously with the Estonian Research Institute for Forestry and Nature Protection and has a Ph.D. in economics, with a specialization in managerial and environmental economics

Eva Kraav, Deputy Permanent State Secretary of the Ministry of Environment of Estonia. Dr. Kraav has also been directly involved in the management of the Estonian Environmental Fund and previously was with the Institute of Economics in the Estonian Academy of Sciences. She has a Ph.D. in Economics, with a specialization in regional and environmental economics.

Kalle Türk, Vice Manager for Ambient Air and Water Protection, City of Tallinn Environmental Board. Mr. Türk graduated from Tartu University with a degree in geological engineering and a specialization in hydrogeology. He was previously with the Environmental Information Centre of the Ministry of Environment of Estonia.

Hungary

Glenn Morris, Project Associate, Harvard Institute for International Development and Senior Environmental Policy Advisor to the government of Hungary. In Hungary, Dr. Morris has directed extensive analyses of Hungary's environmental product charges and collaborated with Hungarian economists and engineers in development of a computable general equilibrium model for assessment of environmental tax policies.

Peter Kovacs has an M.Sc. in Civil Engineering, with a specialization in environmental sanitation, and currently works in the Ministry of Environment and Regional Policy in Hungary. Before joining the Ministry, Mr. Kovacs was an environmental inspector for the Water Quality Department of the Upper Tisza Environmental Inspectorate in Hungary, where he issued permits and fines, conducted water pollution control and supervised treatment facilities.

Jozsef Tiderenczl is a timber industry and environmental protection engineer. He is the director of the Environmental Protection Department of the South Transdanubian Inspectorate in Hungary, where he directs air quality and noise protection programs. Mr. Tiderenczl is also the leader of the NILU Project, which aims to introduce the Norwegian air quality management model in Hungary, and a lecturer at the JPTE Pollack Mihaly Technical College.

Latvia

Janis Brunenieks, Executive Director of the Latvian Environmental Protection Fund. Prior to this position he was with the Economics Department of the Ministry of Environmental Protection and Regional Development in Latvia. He is a lead author of the recently passed Law on Taxation of Natural Resources that is the main focus of the Latvia chapter.

Aija Kozlovska is an environmental consultant to various international organizations working in the Ministry of Environmental Protection and Regional Development in Latvia. She was an assistant with the Harvard Institute for International Development environmental economics and policy project in Latvia during the preparation of this book, and is currently a graduate student at the University of Latvia.

Lithuania

Linas Čekanavičius, Head of the Department of Systems Analysis, Vilnius University, and consultant to the Harvard Institute for International Development. Dr. Čekanavičius has conducted research and consulted on the economic valuation of environmental damages and has been involved in the design of a number of Lithuania's environmental laws and regulations. His research and advising interests are focused on economic instruments for environmental management and the economic valuation of environmental damages.

Daiva Sėmenienė, Chief (on-leave), Economics and Programming Division, Ministry of Environmental Protection and consultant to the Harvard Institute for International Development. Dr. Sėmienė co-authored all of Lithuania's natural resource tax laws, including laws and regulations related to pollution charges. Her research and advising interests currently focus on the use of economic instruments for environmental management and on methods for quantifying and penalizing environmental damage.

Poland

Glen D. Anderson is an environmental consultant based in Raleigh, North Carolina. Dr. Anderson was formerly a project associate with the Harvard Institute for International Development, and from 1993–1996 was senior environmental policy advisor to the Ministry of Environmental Protection, Natural Resources and Forestry in Poland.

Boguslaw Fiedor is Professor of Economics in the Department of Economics, Wroclaw Academy of Economics. Dr. Fiedor is also a consultant to the Harvard Institute for International Development and other organizations.

Romania

Clifford F. Zinnes, Institute Associate, Harvard Institute for International Development. Since late 1993, Dr. Zinnes has been seconded to the Government of Romania as the sustainable development policy advisor to the Council on Coordination, Strategy and Economic Reform. Dr. Zinnes has co-authored many of Romania's environmental laws and also directs a sustainable economic development policy think tank in Romania.

Russia

Anil Markandya is Professor of Economics at Bath University in the United Kingdom, and is Faculty Associate at the Harvard Institute for International Development. From 1994 to April 1996 he was the Senior Environmental Policy Advisor to the Ministry of Environmental Protection in the Russian Federation. He has held advisory positions with most international institutions involved with issues of environmental policy, and has worked closely with the United States Agency for International Development, the European Commission and the government of the United Kingdom. His current research interests include issues of environmental policy and employment, and privatization and the environment.

Michael Kozeltsev is Senior Economics Consultant to the Harvard Institute for International Development Environmental Eonomics and Policy Project in Russia and a consultant to a variety of organizations. He is currently on leave from the Economics Department of Moscow State University.

Slovakia

Thomas H. Owen, Project Associate with the Harvard Institute for International Development and Senior Environmental Policy Advisor to the government of the Slovak Republic. A former provincial government deputy minister in Canada, Dr. Owen's work with the Slovak Ministry of Environment has focused on environmental economic policy and on the systematic application of environmental analytical techniques.

Danka Jassikova has a Ph.D. in chemical engineering. She previously worked as a senior official in the Department of Water Protection in the Ministry of Environment of the Slovak Republic, and currently is an Environmental Management Consultant with the Harvard Institute for International Development.

Jozef Myjavec, Director of the Department of Economic Tools in the Ministry of Environment of the Slovak Republic. Prior to 1991 Mr. Myjavec was Deputy Director of the Bratislava Water Works, responsible for financial policy. His work currently focuses on economic projections, implementation of economic instruments for environmental management, and financial policies for environmental protection.

Preface

The countries of Central and Eastern Europe (CEE) and the former Soviet Union (FSU) began their transition processes with some of the worst environmental policies and problems on the entire globe. Though perhaps the Chernobyl disaster personifies the environmental problems of the CEE and FSU economies most vividly, the evidence is clear that major environmental problems were not isolated or restricted to particular countries, but were endemic to planned economies. For example, as of the late 1980s, dust emissions per thousand dollars of GNP on average were thirteen times higher in the CEE and FSU than in the European Economic Community. Emissions of other air pollutants, as well as wastewater discharges, were two or three times that of the EEC per dollar of output.

These are part of the legacy of planning, and the major reasons for these outcomes are well known. The obsession of communist planners with heavy industry and the associated subsidization of energy go a long way toward explaining the environmental disaster that was exposed when the Berlin Wall came down. For example, at the end of the 1980s the steel production capacity of the Soviet Union was almost twice that of the United States, with an economy one-eighth as large. This pattern of promotion of so-called 'commanding heights' heavy industries was repeated throughout the region and it should be emphasized that these policies were only possible with the support of massive energy subsidies. Low-cost energy in turn not only supported a distorted economic structure, but also encouraged massive inefficiency and excessive air pollution.

Before 1990 it was not meaningful to talk in any serious way about 'integration' of the planned economies into the world economic system. Virtual autarky served to insulate the CEE and FSU from market forces that could have revealed at a much earlier date the region's comparative advantage in much less environmentally destructive industries than steel, cement and the like. Autarky in the form of restrictions on foreign investment also kept newer – and cleaner – technologies out.

Finally, the lack of well-defined property rights simply encouraged waste and carelessness. Without hard budget constraints, prices – whether they were high or low – indeed lost their relevance, because company managers did not bear any penalty for inefficiency or receive any benefit from doing more with less. Similarly, if environmental accidents occurred, state-owned companies were unlikely to be held responsible and therefore did not take adequate precautions to protect populations from environmental accidents.

To a large degree the economic transitions on-going in CEE and FSU countries – combined with crucial democratic reforms – have eliminated the most egregious of these past problems. Throughout the CEE, emissions and damages from major pollutants are down by at least 30 per cent since 1990, and only a portion of those declines can be explained by reductions in output. Countries that have begun the journey to economic recovery are doing so on the basis of more environmentally friendly industrial and economic structures and under much more open rules of governance; clearly, the spread of markets, democratic institutions and citizen participation have been good for the environment. But this is not enough if CEE and FSU countries are to complement the economic growth that started in the early to mid 1990s and which everyone hopes will continue. Better environmental policies that pursue realistic environmental goals, that provide on-going incentives for efficiency, and that attempt to reflect societies' preferences for environmental quality in cost-effective ways, are needed if governments are not to save the planet at the expense of the inhabitants or *vice versa*.

In view of the policies that dominate the West, perhaps the main concern should be the use of expensive and inefficient policy programs that middle-income economies in the CEE and FSU can ill afford. Indeed, evidence is emerging that low – or even negative – cost abatement measures abound in the region. This means that the use of rigid policies that are common in western countries are likely to cause these countries to pass up important opportunities for major reductions in emissions at surprisingly low cost.

Economists talk a lot about the need for more efficient and cost-effective policy regimes, using a variety of economic instruments, but to be effective these tools must be integrated into existing structures at least to some degree. There is therefore a major need to link instruments that are known to accomplish environmental goals at lower cost with the largely administrative environmental management systems that exist throughout the region. One of these instruments is the pollution charge, which has been highlighted since the early 1970s as a particularly useful alternative to so-called 'command and control' regulation. A major goal of this book is to document the emerging experience to integrate this instrument, which often existed but worked ineffectively under communism, into revised regulatory structures. Some aspects of this experience are indeed very positive, but other examples suggest challenges that have yet to be overcome. The book attempts to present preliminary analyses of these systems with a view toward defining what works and what does not. From these detailed examinations of ten countries, insights are derived regarding how pollution charges can usefully be applied in transition economies. The book then distills these insights into a set of lessons in the hope that they will assist policy makers throughout the world who are trying to promote cost-effective and efficient environmental policies.

Jeffrey Sachs
Director, Harvard Institute for International Development
Galen L. Stone Professor of International Trade, Harvard University

Acknowledgements

We would first like to thank the governments and people of Lithuania, Latvia, Estonia and Russia who were kind enough to welcome and allow us to collaborate on the interesting set of issues that has led to this book. Of particular note are the Ministry of Environmental Protection of Lithuania, the Ministry of Environmental Protection and Regional Development of Latvia, the Ministry of Environment of Estonia, and the Committee on Environmental Protection of Russia, where we have served as advisors since 1994. We would especially like to recognize the assistance of Mr Arunas Kundrotas, Secretary of the Ministry of Environmental Protection of Lithuania, without whose active support this book could not have been developed. The continued guidance and collaboration of Ljuba Gornaja, Eva Kraav and Janis Brunenieks was also highly valued.

We would also like to thank the US Agency for International Development for financial support, through the Central and Eastern Europe Environmental Economics and Policy (C4EP) Project and the Newly Independent States Environmental Economics and Policy Project, both implemented by the Harvard Institute for International Development. These projects provided the umbrella under which this book was prepared. The support and assistance of Jim Tarrant and Julie Otterbein are particularly noteworthy.

Most of the ideas developed in this book were first discussed at the Workshop on Implementing Pollution Charges in Transition Economies held in Vilnius, Lithuania, in September 1995. We would like to thank Simona Daugintienė and Jurate Varneckienė for making all arrangements necessary for this workshop, which included representatives from eleven countries in the region along with representatives from the United States and elsewhere.

Finally, we would like to recognize the contributions of our colleagues at the Harvard Institute for International Development. Theo Panayotou, the Director of the International Environment Program, has supported this work from the beginning. Alison Howe, and later Kristen Phelps, organized and managed the production of the manuscript. Without their continued support, this book would probably not have been completed.

Foreword Environmental Policy in the Making: Lessons from Central and Eastern Europe and the Former Soviet Union

One of the paradoxes of the history of environmental policy is the choice of the 'command-and-control' approach by market economies and the use of economic instruments (such as pollution charges and permits) by command economies. The fall of the Berlin Wall revealed the total ineffectiveness of what were no more than accounting devices devoid of any incentive to control pollution. At the same time, developed market economies which earlier thought of costly environmental regulations as a luxury they could afford, began to question the wisdom of rigid regulations as they began to climb the steep part of the pollution abatement cost function: effective they were, affordable they were not. Estimates of the costs of environmental regulations in the United States and Europe range between 2 and 3 per cent of GDP. While, in the aggregate, these costs may still be justified by the benefits, they were neither minimal nor minimized and further cost improvements could only be had at increasingly higher costs. The search for more cost-effective instruments, more consistent with market principles, began in Western market economies more than two decades ago but progress has been slow because of the legacy of regulatory regimes, entrenched environmental bureaucracies and vested interests in the old paradigm.

The collapse of the command economies and the cataclysmic changes that took place in Central and Eastern Europe and the former Soviet Union in the late 1980s and early 1990s uncovered an environmental disaster of similar proportions to the economic disaster; it has also opened the road for radical changes in the conduct of environmental policy. In a climate of fundamental political and economic reforms, of quadrupling of prices, of precipitous decline in economic output and of privatization of large parts of the economy, the prospects of a more cost-effective environmental policy appeared better than elsewhere. The fact that the former command economies were familiar with pollution 'charges' and 'permits' has helped, as did the fact that economic reforms that promote privatization and competition alone began to give these instruments some 'bite'. But even in countries that have changed their economic systems from 'Marx to market' there was no clean slate. Nevertheless, formerly command economies have accomplished more in the past five years in reforming their environmental policies than OECD countries and other market economies, with a few exceptions, over the last 20 years. While there is still a long road to travel, the experiments under way in Central and Eastern Europe and the former Soviet Union hold lessons not only for the transitional economies of the rest of the world but also for market economies facing a regulatory stalemate.

Developing countries that basically copied the Western environmental policy are finding that command-and-control regulations and their end-of-the-pipe focus are not only costly but also ineffective in the absence of the strong legal system and monitoring and enforcement capability. Budgetary pressures, in the form of budget deficits and taxpayers' fatigue, coupled with rising pollution levels and growing environmental awareness, have raised the interest in more cost-effective regulatory policy and in non-conventional sources of revenues for environmental investments. While economic instruments are slowly being introduced in many developing countries they are largely mere add-ons to regulatory instruments, aiming to raise revenues, rather than integral parts of a 'new rules and incentives' system that aims to (a) increase flexibility and compliance, (b) reduce costs, (c) realign incentives, and (d) shape expectations.

The Central and East European and former Soviet Union experience of the past five years, reviewed in this book, holds useful lessons for developing (and developed) countries as to how to develop complementary systems, and how to integrate flexibility, cost effectiveness and tradability in the existing command structure. This volume distils the transitional economy experience into a set of key lessons for designing, implementing and enforcing a combined permit–charge system not through radical replacement but through gradual reform and evolution from what is basically an administered system to a market-driven system. The experience reviewed is short and contains more lessons of 'process' than 'product' since most of the reforms are too new or experimental to draw definite conclusions and include as many lessons of 'how not to' as 'how to'. For example, charges are clearly shown not to provide a strong incentive for pollution control when they are based on self-reporting and the regulators lack the resources for regular unannounced checks.

Similarly, changes are not effective instruments in the presence of soft-budget constraints (for example, state enterprises) and the absence of competition (private or state monopolies). Moreover, setting charge rates in a situation where pollution control costs are poorly understood and environmental benefits are ignored may raise revenues but does not necessarily reduce pollution, much less improve welfare. It is far preferable to recognize and try to accommodate the conflicting interests of enterprises wanting to minimize compliance costs to stay in business and of regulators wanting to maximize revenues for any given level of pollution control to cover the costs of government environmental programmes. Negotiating with enterprises while avoiding the 'regulators' capture', environmental authorities could strike a balance through a charge–credit system that provides the correct incentives at the margin but reduces the total payments made by the enterprise. Allowing enterprises to credit, against their total pollution charge, some portion of their pollution-abatement investments (to the extent that they yield higher return than public investments) reduces both compliance costs and the need for revenues for large public investments. The polluter pays principle and economic efficiency may also be accommodated if the total charge payments are set equal to the total value of environmental damages at the optimum level. These are some of the directions towards which the transitional economy experience and experimentation points for future environmental policy reforms.

Another positive evolution is towards increasing flexibility and tradability of pollution permits, first among sources within the same enterprise then among enterprises and ultimately among sectors (for example, between industry and agriculture or industry and households) within the same airshed. Since under this system the industry retains and invests the expenditures on pollution permits trading, it is a natural evolution from the charge–credit system. Yet another lesson from the environmental policy reforms in transitional economies is the need for stability and predictability of environmental policy. Frequent changes in the 'rules of the game' create policy uncertainty, which is detrimental to investment, especially foreign investment. Gradual implementation and review on a preannounced schedule can help reduce policy uncertainty.

The transitional economies of Eastern Europe and the former Soviet Union are by no means close to an environmental management model that they can confidently implement for the long haul, much less export. Yet their experience and progress along the long road from environmental disaster to recovery under rapidly changing conditions and severe financial constraints, which is recounted and analysed in this book, hold invaluable lessons for others, whether the transitional economies of Asia, or developing countries undergoing less radical reforms. Despite the obvious differences, the similarities are too many and important to ignore:

- like transitional economies, developing countries aspire to (and some have achieved) rapid rates of economic growth and integration into the world economy;
- developing countries have similarly weak environmental institutions and underdeveloped legal systems and lack monitoring and enforcement capability;
- like transitional economies, most developing countries, facing formidable fiscal and financial constraints and pressing needs (for poverty alleviation), can ill afford the high cost of rigid regulatory systems or the fiscal consequences of sizeable environmental investments;
- developing countries share with Central and Eastern Europe and the newly independent states an abundance of win–win investment opportunities and low-cost solutions that can be exploited with the right policy environment;
- finally, developing countries, especially those with high growth rates, are undergoing rapid structural transformation and quick turnover of their capital stock, which offers unique opportunities for institutional change, new rules of the game, and new instruments to shape the expectations of investors and advance pollution prevention over end-of-pipe solutions.

Even Western European countries stand to learn from the experiments taking place across their borders. It is notable that Central and Eastern European countries aspiring to join the European Union, and hence needing to converge to the Union's environmental standards, are searching for more cost-effective instruments that those in use among most EU members. This search transcends the charge systems in use in some Western European countries to tradable permits and other innovative investments used in the United States, New Zealand, Chile and Singapore. Like Western Europe, the US stands to learn from the transitional economies' experiments and experience. For example, tradable pollution permits, being designed in countries such as Kazakhstan, while replicating the positive elements of the US system, are less burdened with a cumbersome registration and approval process and implicit taxation of trading.

In many ways, this is a unique book. It has been written by environmental economists and policy experts who actually are or have been involved in the design and implementation of the policies they describe. It is largely a collaborative work between local and expatriate experts who are reporting on work in progress. Many of the authors have prior OECD and developing country experience in policy design and evaluation. Our hope is that this book will contribute to the debate on the choice, design and implementation of instruments for sustainable development and promote a two-way exchange of experiences and lessons between transitional economies and the rest of the world.

Theodore Panayotou
Director, International Environment Program,
Harvard Institute for International Development

1. Controlling Pollution in Transition Economies: Introduction to the Book and Overview of Economic Concepts

Bruce A. Larson and Randall Bluffstone

1 INTRODUCTION TO THE BOOK

The planned economic systems in Central and Eastern Europe and Russia wasted enormous amounts of natural resources, degraded the environment and damaged the health of citizens (World Bank, 1996; Ahlander, 1994; Environment for Europe, 1994; Goldman, 1985; Sachs, 1995).[1] This waste manifested itself in terms of morbidity and mortality rates that were much higher than in comparable areas (Hertzman, 1993; Environment for Europe, 1994), and in terms of a degradation of land that in many cases made it unsuitable for future use. The initial reaction as these planned systems disintegrated was directed, not surprisingly, at the enormity of these problems (for example, see Slocock, 1992).

The purpose of this book is to examine what the transition economies have been doing since 1989 or 1990 to improve environmental policy systems in order to achieve better environmental and economic outcomes than in the past. Of special interest here is the use of mixed economic and administrative instruments for environmental management, with particular reference to pollution charges and source- and facility-level permits. The book contains chapters that present and analyse the environmental policy systems in Estonia, Latvia, Lithuania, Russia, the Czech Republic, Poland, Slovakia, Hungary, Romania and Bulgaria. Each of these chap-

---

1 This book includes experience from the former Soviet Union for Russia and the three Baltic countries. The Baltic countries are wholly in Europe. Part of Russia is in Europe as well. We mention 'Central and Eastern Europe and Russia' to eliminate any potential political or geographical mishap.

ters was written by individuals who actually are or have been involved in the design, implementation and evaluation of the systems on which they are reporting.[2]

It is hard to imagine more challenging – but potentially promising – circumstances in which to implement environmental policy reforms. Enormous drops in output were the hallmark of at least the first few years of the transition, and these declines have certainly constrained government policies in the region. In the countries considered in this volume gross domestic product (GDP) shrank by anywhere from 18 per cent in Poland and Hungary, to 61 per cent in Lithuania, between 1989 and 1994. The average for European transition countries excluding Russia was 33 per cent, with many economies beginning to grow again by 1993 or 1994 (Fischer et al., 1996).[3]

This transition has offered promise for the environment in part *because* the government failures were so large under previous systems. One result of the enormity of these past government failures has, for example, been that marginal costs of environmental improvements are probably very low in the region, and are certainly much lower than in OECD countries. Outdated technologies and processes were and are easily observable and evidence is emerging that so-called 'win–win' investments that reduce pollution emissions and increase profits may be quite common. From the perspective of enterprises and governments in 1996, these lowest-cost solutions clearly should be chosen before more expensive solutions are considered.

Although the purpose here is not to provide an exhaustive review of the environmental effects of the transition itself, it is clear from the country chapters that another important benefit of the transition has been that dirtier economic sectors have tended to decline by larger percentages than have overall economies. Industry, energy and agriculture have often been particularly hard hit as economies reoriented themselves away from the priorities and methods that were dictated by planned economic systems. As a result, the pollution problems generated by those sectors have also declined sharply. Within just a few years after transformations began, emissions of the most important air pollutants tied to industrial and energy production (for example, SO_2, NO_x, dust) declined by 30 to 60 per cent as the privatization of enterprises and market pricing of inputs and outputs created incentives to reduce waste (Hughes, 1992; Zylicz and Lehoczki, 1993).

Within this fundamentally altered and rapidly changing economic and political environment, it is not surprising that environmental policy has had

2   The country studies are presented roughly from north to south.
3   As reported in World Bank (1996), at least for a subset of the countries included in this book, ground continued to be reclaimed in 1995. The Polish economy, for example, virtually regained its 1989 output level.

to evolve. For example, as late as 1991 framework environmental laws in some countries did not even acknowledge that private sectors existed, but as of 1996 they are an important and in some cases dominant part of most economies. Environmental legislation has therefore been in an almost constant stage of revision since 1990, with many changes being forced by emerging private sectors (or state enterprises operating with harder budget constraints). Regulatory stability is an increasingly important need and it is likely that the rate of policy change will, at the very least, slow down.

Perhaps the most important factor facilitating environmental policy reform has been the large reduction in environmental health threats as emissions of key pollutants have declined. Room has therefore been created for environmental protection ministries to assess their situations, make changes, learn consequences and decide how to change based on their experiences. This breathing space for thoughtful environmental policy reform will, however, probably disappear in the near future as economies pick up. Although it is unlikely that future economic paths will be as pollution and natural resource intensive as the old ones, it is clear that without suitable environmental policy frameworks sustainable economic development will continue to elude transition countries.

To take full advantage of the opportunities for large, low-cost environmental improvements, foreign environmental economists and many environmental economists in the region pushed the notion early on that policies should be revised to offer enterprises both incentives and flexibility to solve environmental problems in the most cost-effective way possible. This, of course, meant avoiding the use of excessively rigid instruments – so-called 'command-and-control' instruments – that have inflated costs in the West. Several environmental economists wrote persuasively on this subject (for example, Dudek et al., 1992; Hughes, 1992; Toman, 1993; Bingham, 1994), and the notion of cost effectiveness was explicitly incorporated into the conventional wisdom surrounding environmental policy for Central and Eastern Europe and the former Soviet Union. For example, in a background note to the Environmental Action Programme for Central and Eastern Europe (EAP) it was concluded that:

> There is a unique opportunity to bring about major environmental improvements in the course of economic transformation and industrial restructuring, and every effort should be made to promote policies which will encourage this process in the most efficient manner possible. (Environment for Europe, 1992)

The use of economic measures, and particularly pollution charges, was and continues to be viewed as an important means for exploiting the opportunities associated with the transition and for achieving more cost-effective environmental policies. Hughes (1992), for example, devoted significant attention to the topic of pollution charge system reform. The background note cited above, as well as the EAP document itself that was adopted at the ministerial conference in 1993, specifically endorsed the increased use of economic instruments in transition countries and recommended that these countries '[build] on existing frameworks of pollution charges' (Environment for Europe, 1994, p. 78).[4]

As demonstrated by the country chapters in this book, all countries to varying degrees include pollution charges levied on estimated or, much less frequently, on directly measured pollution emissions in their environmental policy portfolios. It should be emphasized, however, that this generation of permit and charge systems derives directly from systems that were in place as of 1989 and earlier. The lesson of the transition country experience for the rest of the world is therefore not what happens when one introduces pollution charges in a greenfield situation, but how systems can be revised to try to improve their performance and how to adapt them to important goals, such as membership in the European Union.

Pollution charges in transition countries are also universally integrated with systems of facility- and source-level pollution permits and in some cases are mere add-ons to those regulatory mechanisms. Central and Eastern Europe and Russia therefore provide potentially interesting and important lessons regarding how pollution charges in various forms interact with standard administrative instruments that also were used under communism. How to integrate these systems – and what are the main obstacles in that process – are therefore perhaps the main lessons to be gleaned from the region.[5] This integration is in general not an easy one, and substantial compromises have been made to meld these different approaches. Despite these difficulties, however, policy makers have insisted and are likely to continue to insist that these systems must coexist. As Smith and Zylicz (1994, p. 12) conclude,

> It is unrealistic to expect that environmental taxes can entirely replace regulatory policies in transition economies any more than market mechanisms have replaced regulatory policies in OECD countries. For the foreseeable

---

4   This enthusiasm was not uniform, however. In particular, Smith and Zylicz (1994), reporting on an OECD workshop held in 1993, noted that because of human resource constraints, 'taxes, levied on measured emissions, are unlikely to be worthwhile during the transition phase'.

5   For a discussion of some of the issues with reference to the Czech Republic, see Bluffstone and Farrow (1995).

future, environmental taxes and other market mechanisms should be used to complement regulatory policies.

How to develop such complementary systems is, of course, the key issue, and it is unfortunate for the countries attempting these reforms that other countries do not offer much guidance. In OECD counties, for example, as of 1995 one could count on one hand the number of direct charges on pollutants (Fournier, 1995).[6]

The reasons why pollution charges are of interest in transition countries are quite general. At least as of 1985, the OECD has strongly supported the use of economic instruments to reduce the aggregate cost of environmental management (Opschoor and Vos, 1989). As was discussed by Poitier (1995), part of the reason for this interest is that economic instruments can simplify complex environmental regulations, can provide more cost-effective solutions to environmental problems, can generate finance for environmental projects, and are more consistent with contemporary notions of sustainable development than strictly regulatory approaches. In transition economies these same concerns, particularly cost reduction and creation of environmental finance, are driving environmental policy reforms. A clear difference, however, is the possibility of exploiting low-cost solutions in Central and Eastern Europe and Russia.

To provide a basic framework within which to analyse permit and charge systems, the rest of this chapter provides a brief introduction to the economics of pollution control, focusing on the assumptions that drive the economic analysis. It is also emphasized that several of the more important assumptions are not likely to be valid in transition economies. Chapter 2 provides a concise overview of the key features of pollution permit and charge systems in the region, with particular reference to commonalities in structure that exist. The authors then introduce and discuss several important design issues that are useful for evaluating the nine country chapters that immediately follow. As was already mentioned, the country-specific chapters focus on details of developing, implementing and evaluating environmental policies in transition countries. This book concludes by distilling this experience into a set of key lessons for developing, implementing and enforcing combined permit and charge systems.

---

6 Comparing the amount of information available on the use of tradable emission rights with that for pollution charges, for example, shows clearly that much more is known about how tradable permits operate because there have been several trading programmes in the United States and substantial analytical work has been conducted. Similar design-based analysis has not occurred in the case of pollution charges and discussions have therefore tended to be largely theoretical.

2 THE ECONOMICS OF POLLUTION CONTROL IN TRANSITION ECONOMIES

The economic theory of pollution control is well developed, and can be reviewed in a variety of texts.[7] This section, therefore, merely tries to make explicit certain elements of this theory that are relevant to the discussion of pollution charges in transition economies. The main purpose is to provide a unified framework from which the country-specific approaches can be understood, contrasted and critiqued. To set out this framework quite explicitly, a list of key assumptions is presented which we think has a bearing on the implementation of these systems in many countries. By examining these assumptions, it is possible to begin to understand the difference between a standard model of pollution control and the actual situations that exist in the region.

To begin, the terminology related to pollution charges can be confusing, and terms like taxes, charges, fees, fines and penalties are often used at the same time and interchangeably in English. For reference, we follow Tietenberg (1992, p. 372), who writes that 'a [pollution] charge is a fee, collected by the government levied on each unit of pollution emitted into the air or water'. In many cases pollution charge rates differ depending on whether total pollution is above or below firm-level pollution standards (also called 'limits' here) specified by environmental authorities. These higher charge rates do not, however, imply that laws are breached; they are levied simply for not achieving the pollution levels envisioned as part of overall programmes of environmental management. In this book such above-limit payments will be called 'penalties', and are distinct from 'fines' which are payments for actually breaking laws or conditions of permits.[8]

2.1 Assumptions and the Basic Framework

This section makes explicit the assumptions underlying functions that determine pollution levels and if these levels are socially correct. On the demand side, the demand for pollution by firms is analogous to any other demand function because generating pollution provides benefits to firms

---

7   For example, Tietenberg (1992), Baumol and Oates (1988) and Mills and Graves (1986) provide presentations that discuss the potential usefulness of economic instruments. There is, of course, a large literature on instrument choice. A partial list includes Bohm and Russell (1985); Milliman and Prince (1989); Weitzman (1974); Roberts and Spence (1976); Cropper and Oates (1992); Hahn (1989); Hahn and Stavins (1992); Tietenberg (1980); Hahn and Hester (1989); Howe (1994); Stavins (1995).

8   A company might pay a fine, for example, for burning solid fuels in a boiler when its permit only allows the burning of liquid fuels.

in the same way that labour and energy generate benefits. Reducing pollution reduces these benefits; this reduction in benefits to the economy is the cost of reducing pollution – abatement costs. On the supply side, damage from pollution defines social costs of pollution (and the benefits of reducing pollution).

Seven intuitive assumptions are necessary to generate demand functions for pollution and to define an optimal supply of environmental services:

A. pollution is generated somehow by a set of production processes;
B. enterprises are able to measure pollution without excessive costs either through direct monitoring or indirect estimation methods;
C. enterprises understand very well the relationship between production and pollution;
D. enterprises try to optimize some objective function;
E. key markets function well, particularly financial capital markets;
F. pollution creates social costs through environmental damages that can be translated into the same units (for example, monetary) as the benefits of pollution; and
G. environmental authorities can observe pollution without excessive costs either through direct monitoring or through indirect estimation methods.

Assumptions A to D are needed to make the link between inputs, outputs and emissions that define the demand for pollution in an economy. Assumption E allows enterprises to finance a larger range of pollution-control options, which reduces the demand for pollution as well as making it more responsive to pollution charge levels. Assumption F adds the cost element and allows us to define a social cost function (environmental damages), as well as a marginal social cost function that is the social willingness to allow pollution (the pollution supply function). Assumption G allows enforcement of some chosen environmental policy approach, and in important respects defines the ability of regulators to achieve or approximate ideal policies.

The country chapters in general emphasize that all of these assumptions do not strictly hold. While Assumption A is obvious for the most part, but not always, it is far from clear that Assumption B holds in many countries included in this book. The lack of pollution monitoring equipment and perhaps outdated estimation methods means that enterprise knowledge of the link between *actual* pollution and *expected* pollution can be quite loose. Companies may want to do the right thing, but they do not themselves (at least now) know their emissions. In the past, it is well known that environmental laws often went unenforced, in which case it was less important to estimate/monitor pollution levels well. The large economic changes noted above have also brought changes in products, inputs, processes and management. Combined with a continued deterioration of capital in many cases, it is not at all clear that the link between production and pollution is well known to enterprises. As Smith and Zylicz (1994, p. 8) note:

> Enterprises begin from a low level of awareness of the various technological options for pollution control, and their associated costs and benefits. This is partly a result of the lack of past contact between western suppliers of pollution control and other technologies, and individual enterprise managers in Central and Eastern Europe.

Although privatization, the elimination of state subsidies and improved enforcement certainly provide incentives for learning, it is likely at least at present that Assumption C often does not hold. For the future, however, through a combination of better market signals and enforced environmental policy, it is likely that this problem will be reduced over time.[9]

Assumption D states that enterprises try to optimize 'something', but this may not necessarily imply a maximization of profits or a minimization of costs. While clear market incentives exist for private enterprise managers to give profits a prominent place in their objective functions, such is not necessarily the case for many state-owned enterprises (some of which will be privatized in the future). Under such circumstances it is not clear exactly what is being maximized. As long as 'something' is being optimized, it is possible to determine some form of pollution demand function.

In order truly to be able to choose their optimal level of emissions, enterprises not only must have full information, but they must also have access to complementary markets – particularly financial capital markets.

---

9   This poor understanding of pollution-control option has created many opportunities for very low-cost pollution-control activities. Work by the World Environment Center, for example, suggests that investment opportunities benefiting both the environment and companies' bottom lines certainly are available and are probably more numerous than is generally recognized. While reducing emissions of various pollutants by significant amounts, selected investments in Poland ranging from $800 to more than $600,000, uniformly paid back in less than one year, and in many cases payback periods were less than one month. Similar results were obtained in the three Baltic countries (World Environment Center, 1995a; World Environment Center, 1995b), as well as by companies in Lithuania that made investments without outside help (Bluffstone and Varneckienë, 1996). These opportunities represented – and there is no reason to believe that as of 1996 they do not still represent – low-cost investments made by those who understand the best ways to improve production processes. Such cases, however, also beg the question whether other opportunities exist that simply require more nudging from a policy environment offering enterprises incentives and the flexibility to solve environmental problems.

If financing is for some reason artificially restricted for the classes of investment projects that are linked to pollution reductions, enterprises' choices are constrained, which makes the demand for pollution (control costs) higher and less responsive to price signals.[10] As is discussed throughout this book, Assumption E is at the very least under serious pressure, and often simply does not hold, either because capital markets are not sufficiently developed or because markets are simply too thin to include financing of environmental projects. To address this market failure, in most countries systems of local and national environmental funds have been developed, and in some cases (for example, Poland) these funds are very large. Questions exist about the efficacy of these funds, however, and substantial work has gone into creating guidelines that lay out clearly the major elements of successful environmental funds (Lehoczki and Peszko, 1994, 1995; Lehoczki and Morris, 1994; Lovei, 1994).

Of course, interest in reducing pollution is some function of regulators' abilities to monitor and enforce (Assumption G). There are, however, substantial questions regarding the capabilities of environmental authorities to monitor and enforce permit requirements and to collect the correct amounts of charges. Continuous emissions monitoring is the exception everywhere in the region and environmental staff levels, usually a local arm of the national level, are generally insufficient to allow frequent checking of enterprises' emissions reports. These monitoring and enforcement problems alone make one question whether pollution charges are right for Central and Eastern Europe and Russia. As Mills and Graves (1986) correctly point out, monitoring and enforcement problems have a negative impact on system performance whether pollution charges or direct regulation is used.

In sum, assumptions A to E imply that an enterprise creates pollution, knows what it is, and is able to choose the level that optimizes its objective function. For the case where enterprises are not regulated, they can freely choose the level of pollution Q that best meets their needs. Mathematically, the enterprise solves:

$$\max_Q B(Q) \Rightarrow MB(Q_0) = 0 \tag{1.1}$$

where the function $B(Q)$ is the enterprise's net benefits from pollution generation, $MB(Q)$ represents the marginal benefits of pollution. The function $MB = MB(Q)$ is the enterprise's demand for pollution in the same way that any demand schedule depicts the marginal benefits of an

10   In microeconomic theory, this idea is analogous to the difference between a short-run marginal cost function and a long-run marginal cost function.

additional unit of a commodity consumed. The first-order condition $MB(Q_0)$ shows that total benefits are maximized at pollution level Q_0, where the marginal benefits equal zero.

The term $MB(Q)$ can also be thought of as the marginal pollution abatement cost schedule, because $MB(Q)$ shows how much the enterprise must give up to reduce pollution by one unit. For example, for a very simple case, if pollution is directly related to output in fixed proportions, say as Output $= A*Q$, where $1/A$ is the pollution intensity of output, the marginal cost of abating one unit of pollution is equal to the benefits (for example, profits) lost from producing $1/A$ units of output.

However, allowing substitutability of inputs and perhaps greater efficiency in the use of certain inputs in production, an additional unit of pollution can be reduced for less than $1/A$ units of output. Such methods might include some combination of production decreases, substitution among inputs, modifications to outputs or end-of-pipe controls. It needs to be emphasized that these marginal abatement costs are the *least-cost* combination of methods for reducing pollution, and this least-cost approach can be very different from engineering cost estimates that are often based exclusively on technological solutions – particularly end-of-pipe methods – that hold output constant. In the transition country context, where substitution possibilities are not fully exploited, however, such formulaic estimates are likely to be inappropriate.

2.2 Instruments to Achieve the Socially Optimal Pollution Level

Now assume policy makers are concerned with the socially 'efficient' level of pollution defined as the level of pollution that maximizes net social benefits from pollution control.[11] Social efficiency takes account of the damages of pollution, which by Assumption F are assumed to be known. Assuming that these benefits and costs can be translated into the same units (for example, monetary units), society's problem becomes:

$$\max_Q B(Q) - D(Q) \Rightarrow MB(Q^*) = MED(Q^*) \qquad (1.2)$$

where MED are the marginal environmental damages of pollution. The efficient level of pollution – the Pareto-efficient level – in this case Q^* occurs where the marginal benefits of pollution equal the marginal social costs of pollution.

---

11   If for some reason society knows the level of pollution it wants, then the planner's strategy should be to minimize the costs of achieving a given level of pollution. Of course, it may also be incorrect to assume that society agrees on the 'right' level of pollution.

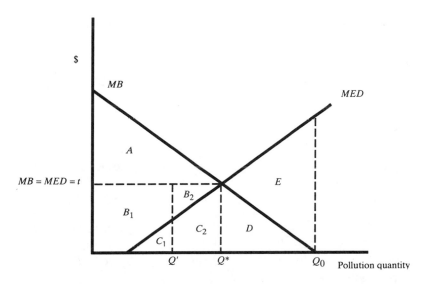

Figure 1.1 A basic economic framework for pollution-control policy

Figure 1.1 summarizes these relationships between marginal benefits (*MB*) and marginal environmental damages (*MED*). For simplicity, this graph presents the world as if it is made up of one firm. The discussion based on this graph will therefore be in terms of 'the firm', but when necessary the implications of extending the model to the multifirm case will also be noted.

By definition, the socially optimal level of pollution balances the benefits and the costs of pollution, but as summarized in Table 1.1 the optimal solution, Q^*, can be reached in several ways. The first approach, using a performance standard, allows the firm to generate Q^* units of pollution. For the multifirm case this total pollution load would need to be *optimally* divided in a least-cost way between all enterprises for a social optimum to be achieved. A second approach, using a pollution charge, makes the firm pay a charge, $t = MB(Q^*) = MED(Q^*)$, on all emissions.[12] A third approach, also using a pollution charge, makes the enterprise pay a charge, $t = MB(Q^*) = MED(Q^*)$, on all units of pollution between Q' and Q^*, but no charge on pollution below Q'. In the multifirm case, each enterprise receives its own Q' that can be thought of as a permitted limit. To achieve a true social optimum, though, all firms must face the same marginal incentive (that is, be on the same step of charges).

---

12 For a uniform pollution charge, expanding to the multifirm case makes little difference.

*Table 1.1   Pollution benefits, damages and abatement costs with differ-
ent charge policies*

	Unregulated (Q_0)	Regulated optimum (Q^*)	Optimal charge (Q^*)	Two-step charge (Q^*)
1. Private benefits	$A + B + C + D$	$A + B + C$	A	$A + 1B$
2. Charges paid	0	0	$B + C$	C
3. Environmental damage	$C + D + E$	C	C	C
4. Net social benefits $(1 + 2 - 3)$	$A + B - E$	$A + B$	$A + B$	$A + B$
Private costs of pollution (abatement and charges)	0	D	$B + C + D$	$C + D$

Note: From Figure 1.1, $B1 + B2 = B$ and $C1 + C2 = C$. And, for the two-step charge
system, the quantity Q' is chosen so that $C1 = B2$.

Although all three of these approaches can in principle achieve the
socially desirable level of pollution, Q^*, the main point we would like to
emphasize is that each policy has very different distributional implications attached to it. In Figure 1.1 and Table 1.1, for the base case of the
unregulated firm, the pollution level Q_0 is chosen and the firm obtains
total benefits of $A + B + C + D$. Social damages equal $C + D + E$. As
a result, net social benefits with an unregulated firm equal $A + B - E$.[13]
    The policy approach labelled performance standard just implies that the
environmental regulator directly chooses the right level of pollution Q^*.[14]
With this approach, net social benefits increase to $A + B$. Under this policy
a polluter pays abatement costs of D to reduce pollution from Q_0 to Q^*.
Such a policy represents an application of the *polluter pays principle* recommended by the OECD because under a performance standard

13   In this discussion, $B = B1 + B2$ and $C = C1 + C2$ in Figure 1.1.
14   It is important to emphasize that this performance standard is very different from the
strict notion of 'command and control' (CAC). CAC regulations – at least in their most rigid
form – specify technological standards for pollution control, not pollution quantities. A per-
formnce standard offers substantially more flexibility than technological standards. For the
most part, performance standards are used by the transition economies included in this book.

enterprises are responsible for paying abatement costs necessary to reduce emissions to socially desirable levels (Pezzey, 1988).[15]

Under a uniform pollution charge policy, the firm voluntarily chooses emissions of Q^*, because profit maximization will lead the firm to choose the level of emissions such that $MB = t$. Net social benefits are equal to $A + B$, which is the same as with the optimal performance standard. The distributional impacts of a uniform charge policy are quite different from that of a performance standard, however, because besides paying abatement costs equal to D, a polluter also pays $B + C$ in total pollution charges. The total pollution charge can then be divided into two types of payments. The amount C equals the value of the damages created by pollution level Q^*. In some sense, C is compensation that an enterprise pays for the use of the environment. The amount B is an extra charge – essentially a tax – paid to the government because of the instrument chosen. *Under a uniform pollution charge, policy enterprises pay more than the value of their environmental damages,* and they are, of course, worse off than with a performance standard.[16]

Not surprisingly, enterprises and often environmental authorities are not too interested in using uniform charges at high enough levels to achieve socially optimal emissions because of the potentially negative financial impacts on enterprises. In many transition countries, where establishing viable market economies is the number one concern, mandating substantial transfers to the public sector simply as a tactic for internalizing environmental externalities is quite questionable. Notice also that this uniform charge approach is not clearly consistent with the original idea of a *polluter pays* approach to pollution control because the firm pays more than just the costs of pollution abatement.

A conceptually simple way to solve this distributional problem is to implement a two-step charge where an enterprise pays a charge $t = MB(Q^*) = MED(Q^*)$ on all units of pollution between Q' and Q^*, but no charge on pollution below Q'.[17] In Figure 1.1, the level Q' was chosen so that $C1 = B2$. In this case with a two-step charge system the enterprise pays total charges just equal to the environmental damages it creates. More on this issue is discussed in Chapter 2.

---

15 While not too important for this discussion, there are several versions of 'polluter pays' and, at least in the authors' experience, the words are often used without a precise meaning.
16 This fact is a standard byproduct of competitive markets as well. For example, we pay the market price for our final unit of a commodity, but receive consumer surplus from all other units.
17 There are many variations on this approach. The important point to emphasize is that the marginal incentive must be the same as under a uniform charge. Of course, we are not advocating that this idea is directly implementable in practice.

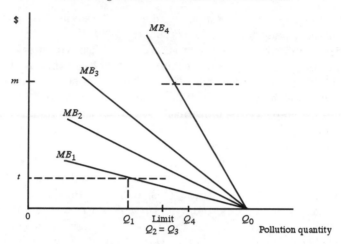

Figure 1.2 A two-step pollution charge policy

2.3 Charge Structures in Transition Economies

All of the transition countries included in this book use some variation of a two-step charge system, where the quantity Q' in Figure 1.1 is a quantity limit specified in a pollution permit. Figure 1.2 presents the possible results of implementing a two-step charge structure for four different enterprises. Q' is the permitted limit, t is a base pollution charge rate (for example, $/ton) charged on pollution levels up to the quantity Q', and m is a penalty levied on all quantities above Q'. In some countries t is zero and in some countries it is positive. The charge m is positive in all countries and typically a simple multiple of the within-limit charge rate.

The marginal pollution benefit functions MB_1, MB_2, MB_3 and MB_4 show four possible relationships between charge levels, limits in permits and marginal benefits of pollution. Although typically standards are tailored to enterprises' particular circumstances, to make a graphical representation possible, a uniform standard for all four firms is assumed first.[18]

18 Marginal benefits of pollution differ for at least four main reasons. First, enterprises produce different types of products that either use different amounts of polluting inputs or generate less pollution in the production process. Second, enterprises' technologies differ, creating differences in the use of inputs or use the same inputs more or less efficiently. Third, some enterprises may have already invested in end-of-pipe controls. Because these costs are sunk, one ton of abatement comes at a lower short-run marginal cost than for firms that have no control equipment. Finally, enterprises may have exactly the same technology, but some may simply be better managed so that the greater efficiency is obtained for the same technology.

These differing marginal benefit functions determine how enterprises respond to the use of pollution charges.

For the enterprise with marginal benefits MB_1, $MB_1 = t$ at pollution level Q_1, and the charge creates an incentive to pollute under the limit Q'. Second, for enterprises with marginal benefits MB_2 and MB_3, $t < MB_2 < MB_3$ m at pollution level Q' so that the charge provides an incentive to pollute at the limit $Q' = Q_2 = Q_3$. In this case, the two enterprises try to avoid paying the penalty rate m and emissions are limited to the uniform standard. Finally, for the enterprise with marginal benefits MB_4, $MB_4 = m$ at pollution level Q_4 and the penalty rate m encourages the enterprise to reduce pollution from the unregulated level Q_0, but it will not voluntarily choose to stay within the limit Q'.

Figure 1.2 is useful for reemphasizing the importance of Assumptions A–D. An enterprise needs to be able to monitor or estimate its pollution emissions and must also know its marginal benefit function, otherwise it is very difficult for pollution charges to create any incentives for least-cost pollution abatement. As was already mentioned, both of these assumptions, monitoring and knowledge of marginal abatement costs, are in question in many transition countries.

Regarding revenues collected from a two-step charge system, more revenues are collected when the marginal benefit function is relatively steep (MB_4 compared to MB_1). In effect, pollution-control policy costs more in economies with marginal benefit functions such as MB_4, and the government therefore collects more revenues. One of the most important advantages of a uniform charge system is that it is a 'cost-effective' policy; meaning that any level of pollution control is achieved at the least cost to all enterprises. Cost effectiveness is assured when marginal benefits of pollution are equated across all enterprises at their chosen levels of pollution. Thus, a potentially important critique of a two-step charge structure as implemented in the transition economies included in this book is that in Figure 1.2 at the pollution quantities Q_1, Q_2, Q_3 and Q_4, the marginal benefits of all four firms are different.

The practical importance of this cost-effectiveness issue depends on country-specific conditions regarding the standards specified in permits and the level of charge rates. From Figure 1.2, if all enterprises are polluting within their pollution limits they are all paying the same base charge rate of $t = MB$ and the system is cost effective. Similarly, if all enterprises are above their limits, they all pay the higher rate $m = MB$ and the system is cost effective. Figure 1.3 develops the idea of cost effectiveness more specifically for enterprises with the marginal benefit functions MB_2 and MB_3 from Figure 1.2. The vertical axis on the left shows the marginal benefits and charge levels for the enterprise with MB_2, and the vertical axis on the right contains marginal benefits and charge levels for the enterprise with MB_3. The horizontal axis read from left to right denotes pollution quantities for enterprise 2, and the horizontal axis read from right to left denotes pollution quantities for enterprise 3. Since each enterprise pollutes at the limit Q' in Figure 1.2, the total quantity on the horizontal axis in Figure 1.3 is twice the limit ($2Q'$). At the limit Q', $t < MB_2 < MB_3 < m$, and total costs of pollution control are $A + B + C$ (A to enterprise 2 and $B + C$ to enterprise 3). From Figure 1.3, if marginal benefits were equated at pollution quantity Q for enterprise 2 and $2Q' - Q$ for enterprise 3, total costs of pollution control would be $A + B$. Thus, the quantity C represents how 'cost ineffective' this uniform limit is.

It needs to be emphasized, however, that it is differences in marginal benefits evaluated at actual pollution levels (for example, the limit in Figure 1.3) that determines if a policy is or is not cost effective, but it does not determine just how much additional cost the policy imposes on enterprises. In effect, marginal benefits must be substantially different so that the area 'C' in Figure 1.3 is large. If, however, MB_2' instead of MB_2 is the second enterprise marginal benefit function, the loss is only C_2; although the limit Q' enforced by the two-level charge system is not strictly cost effective, it is not too far off. Without detailed analyses of marginal benefit functions – which do not exist for most parts of the world – it is not clear how much additional cost is imposed on enterprises by using a two-step charge system. These issues are discussed further in Chapter 2 and the country-specific chapters.[19]

2.4   Relaxing the Assumptions of Perfect Information: Implications for Policy

To an important degree the country chapters in this book are about how and why the standard assumptions of the theory of pollution control cannot reasonably be thought to hold in transition economies, and why this divergence is important. Perhaps the most important of these assumptions deal with information. Assumptions B, C, E and F all focus on the need by regulators or polluters for certainty in order for the use of instruments such as pollution charges to translate neatly into efficient

---

19   Figure 1.3 also shows the potential benefits of allowing enterprises to trade pollution quantities. At the limit Q', $MB_3 > MB_2$. Enterprise 3 would be willing to pay enterprise 2 up to MB_3 for enterprise 2 to reduce pollution, rather than actually having to do it itself. At the same time, enterprise 3 would have to pay at least MB_2 to convince enterprise 2 to agree to reduce pollution one more unit.

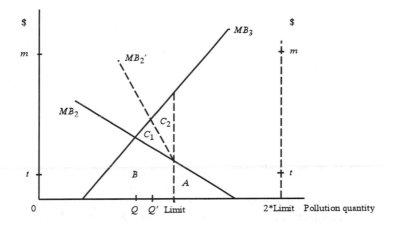

Figure 1.3 Cost effectiveness and a two-step charge policy

actual – or at least expected – levels of pollution. Somewhat roughly speaking, we can say that these assumptions deal with the following:

- the distinction between actual and reported emissions;
- the position of marginal benefit functions for individual firms and for groups of polluters;
- the position of marginal damage functions;
- the effects of environmental investments stimulated by policy instruments on total emissions.

As was already mentioned, monitoring and enforcement is a major problem in Central and Eastern Europe and Russia. For many pollutants included in pollution charge systems continuous monitoring equipment *per se* simply does not exist and estimation techniques may be very rough. This technological gap is not, however, typically filled by more frequent spot checking. In Lithuania, for example, enterprises can expect to be checked no more often than quarterly, and it is more likely that they will only be visited annually. Spotty monitoring creates opportunities for misrepresentation, and as pollution charge rates rise, incentives to evade charges also rise. A particular challenge, therefore, is to create systems with built-in incentives for accurate reporting, even if estimates done with the best of intentions may not be precise. Most countries, for example, already have fines to deter misrepresentation, and Poland denies access to its extensive system of environmental funds if all pollution charges are not paid.

From the perspective of an enterprise that reports accurately, the main issue is perhaps that it is likely to have only a very vague knowledge of the abatement options that exist. Indeed, solutions to environmental problems typically are very plant and pollutant specific, making generic solutions unavailable, and investments often affect more than one pollutant simultaneously. This feature alone vastly complicates the picture and makes very difficult a one-to-one mapping between a unit of investment and a unit of pollution reduction over time. A further complication is that most environmental projects are actually resource-conservation projects that reduce production costs, which makes abatement cost estimates difficult, often simply sidelining the environmental question.

Information is very costly in the countries in transition, and firms are unlikely to know the cheapest options for reducing pollution. Until information systems are better developed, companies must operate with the knowledge that exists and only afterwards will they learn that the pollution levels are higher or lower than expected, or that cheaper options were available. This issue is usually ignored, or at best addressed by assuming enterprises operate according to 'expected' marginal benefit functions, which would be, for example, some combination of the four marginal benefit functions in Figure 1.2. An enterprise therefore chooses its emissions where expected marginal benefits equal pollution charge rates, but it pays based on the level of pollution that emerges from the actual marginal benefit function.[20] Enterprises are not reassured when they make the right decision *ex ante* in terms of expected values, but then pay a substantially higher level of charges than planned *ex post*. For the purposes of this book, and the applications of charge policies in transition economies, this point is not emphasized enough in introductory textbooks.

Uncertainty also has some implications for the usefulness of choosing a policy approach based on limits (Q' in Figure 1.3) or a single charge level. Based on the analysis of Weitzman (1974) and as explained in simple terms in Tietenberg (1992), a limits approach is preferred over a charge when the marginal environmental damage function is 'steeper' in Figure 1.1 than marginal benefit of pollution function. And, in contrast, a charge approach is preferred over a limit when marginal benefits of

20   In terms of Figure 1.2, for example, this could be interpreted as follows. Due to uncertainty, an enterprise does not know before the fact if its marginal benefit function is MB_1, MB_2, MB_3 or MB_4. If the enterprise incurs some cost of X on the horizontal axis between the base rate t and the penalty rate m, then three pollution outcomes are possible: below Q_1 if MB_1 is the final benefit function; equal to Q' if MB_2 or MB_3 is the final benefit function; or above Q_4 if MB_4 is the final benefit function. In this situation, the enterprise undertakes a gamble; it undertakes actions for some cost, and then waits to learn the final outcome in terms of pollution levels. When it learns its final pollution level, it then learns for that period essentially which marginal benefit function it was working with.

pollution are steep compared to marginal environmental damages. The potentially complementary link between charges and limits to handle uncertainty problems (Roberts and Spence, 1976) is discussed more in Chapter 2, the country chapters and the concluding chapter.

Uncertainty about marginal benefits of pollution/abatement costs has clear policy implications, because estimates of the responses to possible pollution charge rates must be made before implementing particular policies. But the response of enterprises can, of course, only be determined by knowing or at least estimating some marginal benefit function for the relevant polluters. This is not easy information to obtain, and the paucity of studies covering heterogeneous firms and more than one or two pollutants testifies to the difficulty of estimating marginal benefit functions. Often the best that can be done is to compute an *average* abatement cost. Because the actual process of abatement option choice is so complicated, often these estimates are also limited to end-of-pipe controls for which engineering estimates are available.[21]

In some circumstances, it may be reasonable to assume that marginal environmental damages are constant over a relevant range of emissions. If this simplification can be made, which makes the analyst's job much easier, marginal environmental damages in Figure 1.1 would be constant. This of course simplifies estimating marginal benefits, which are a single constant value and not a function of emissions. With a charge set equal to this estimated damage, firms can be expected to respond by maximizing the value of their objective functions and regulators will sleep well knowing the resulting emissions levels will be efficient. One potential problem is that analysts will not be able to predict what those emissions will be, and policy makers will therefore be unlikely to endorse such approaches. In general, therefore, for pollution charges to fulfil anything more than a vaguely defined revenue generation function, estimates of marginal pollution benefit functions must be made.[22]

A final topic to discuss is the effect of charges on incentives for investment in pollution control, and the effects of those investments on total pollution emissions. If Assumptions A through E more or less hold, demand functions for pollution and pollution reduction will exist. A charge policy will then create incentives for enterprises to invest to reduce the

21 This was the approach taken, for example, in Sweden when a charge on nitrogen oxide emissions emitted by power plants was instituted. End-of-pipe abatement cost estimates were made for several companies and an average was taken to set the charge rate. The resulting level of pollution reduction was, however, much greater than expected, suggesting that marginal benefits had been overestimated.
22 To accurately estimate revenues, estimates of marginal pollution benefits must also be made.

short-run marginal benefits of pollution by using better processes, management and technologies. In the simple model presented so far, however, a one- or two-step charge always imposes higher costs on firms than a pure pollution standard. The implication of this higher cost is that, at least in principle, the benefits of investment and innovation are larger with a charge than with a performance standard.[23] Whether any investment actually takes place, however, depends on whether the investment costs are less than the benefits of pollution evaluated at the optimum; just because a charge creates incentives does not guarantee that they will be sufficient for enterprises to actually invest. Cause for concern is particularly justified given the often thin credit markets in transition economies, which increases investment costs. With imperfect credit markets, charges may not create incentives to invest and indeed could reduce enterprises' abilities to invest by siphoning off investable funds.

To consider more carefully the relationship between charges, investment incentives and changes in emissions, it is necessary to ask explicitly how pollution is generated, how investment alters emissions, and how such investments alter enterprise-level marginal benefit functions. The standard notion of investment in pollution control is illustrated in Figure 1.4 by two marginal benefit functions, MB(old) and MB(new). The usual story goes that investments that reduce the pollution intensity of production (for example, changes in product mix, production processes and installation of end-of-pipe controls) alter the short-run relationship between costs of pollution reduction and emissions levels such that the marginal benefit function pivots downward; short-run marginal pollution abatement costs are therefore everywhere lower after investments are made.[24] MB(old) is the original case, and after investment MB(new) is the new situation. The assumption is that investment in pollution control reduces marginal benefits for all units of pollution.

In general, though, it is not clear if investments in pollution control always imply decreases in marginal benefits functions, and this means that regulators face substantial uncertainty regarding the effects of policy instruments. In fact, it is not too difficult to create examples where this situation does not hold. While the details are presented in the appendix

23   See Tietenberg (1992) and Pezzey (1988) for a more complete discussion of incentives for pollution control under pollution charges and standards in the context of this type of simple model.
24   The focus here is on the short run, because for the foreseeable future in addition to issues of investment, firms in transition economies are also likely to be interested in short-run optimization questions related to better housekeeping, better process controls and fuel switching. Focusing only on long-run marginal abatement cost curves would obscure these relatively lower-cost options.

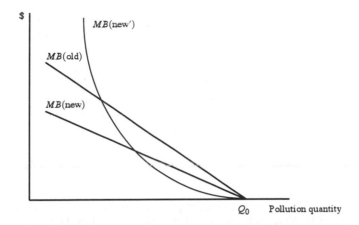

Figure 1.4 Marginal benefits and investment in pollution control

to this chapter, a basic result is also summarized in Figure 1.4. In Figure 1.4, rather than investment changing the marginal benefits function from *MB*(old) to *MB*(new), it could instead result in something like *MB*(new'). This might occur if an enterprise's output determines pollution levels and investments primarily reduce the pollution intensity of output. Rather than a uniform decrease in marginal benefits, *MB*(new') is below *MB*(old) for some pollution levels, but is above *MB*(old) at other levels, and therefore for some charge levels enterprises would reduce pollution more if no investment in pollution control occurs.

The logic is that investment in pollution control reduces the pollution intensity of output, which reduces the marginal benefits function. This change also, however, reduces the weight of pollution charges in the cost structure and as a percentage of the output price, so the net output price (that is, per unit profit) increases. With higher net output prices, enterprises produce more. If production increases enough, pollution levels could also increase even though the pollution intensity of output decreases.

In the appendix, a second situation is also considered where pollution is generated by some input or group of inputs. In this case, investment in pollution control is equivalent to investment in efficiency improvements, so that the same level of output can be produced with fewer polluting inputs. In the appendix it is shown why investments in efficiency in the use of polluting inputs can actually increase the marginal benefits function so that *MB*(new) is always above *MB*(old).

The purpose of Figure 1.4 and the discussion in the appendix is to show that the effects of investment on pollution levels depend on several factors, including: (1) the economic relationships between production technologies; (2) how pollution is generated; (3) how investment in pollution control alters these relationships; and (4) pollution charge levels. As will been seen in the country chapters, there is a general hope that pollution charges will spur investments that will result in lower total emissions. The theory focusing on such issues, however, does not provide automatic reasons why this will certainly occur. Practical experience also does not, at least as yet, provide clear empirical results on this question.[25]

REFERENCES

Ahlander, A.S. (1994), *Environmental Problems in the Shortage Economy: The Legacy of Soviet Environmental Policy,* Aldershot: Edward Elgar.

Baumol, W.J. and W.E. Oates (1988), *The Theory of Environmental Policy,* 2nd edn., Cambridge: Cambridge University Press.

Bingham, T. (1994), 'Options for Utilizing Residuals Charges in the Czech Republic', Paper Prepared for the Ministry of Environment of the Czech Republic, December.

Bluffstone, R. and S. Farrow (1995), 'Adapting Economic Instruments Within Existing Administrative Systems: The Case of Air Pollution in Northern Bohemia', Mimeo.

Bluffstone, R. and J. Varneckienë (1966), 'Average Annualized Abatement Costs of Pollution Control Investments in Lithuania', Mimeo.

Bohm, P. and C.S. Russell (1985), 'Comparative Analysis of Alternative Policy Instruments', in A.V. Kneese and J.L. Sweeney (eds), *Handbook of Natural Resource and Energy Economics*, Vol. I, Amsterdam: North-Holland, pp. 395–460.

Cropper, Maureen L. and W.E. Oates (1992), 'Environmental Economics: A Survey', *Journal of Economic Literature,* **30**: 675–740.

Dudek, D.J., R.B. Stewart and J.B. Wiener (1992), 'Environmental Policy for Eastern Europe: Technology-Based Versus Market-Based Approaches', *Columbia Journal of Environmental Law,* **17** (1): 1–52.

Environment for Europe (1994), *Environmental Action Programme for Central and Eastern Europe* (abridged version), 31 March.

Environment for Europe (1992), *Environmental Action Programme for Central and Eastern Europe* (background note), 6 April.

25   Policy uncertainty is another type of uncertainty that has only begun to be analysed in the economics literature (for example, see Larson and Frisvold, 1996), but which is common to all transition economies. The basic idea is, because the very nature of transition economies is change, enterprises today may not have a clear idea of what environmental policy will be tomorrow. Real charge rates have been adjusted over time, permitting procedures change, effluent and ambient standards are revised, and enforcement mechanisms altered. For the case where pollution is generated directly by an input or group of inputs, a basic result from Larson and Frisvold can be summarized as follows: policy uncertainty creates incentives for enterprises to invest in technologies, which improves their ability to respond to pollution charges in the future (that is, reduces the slope of the marginal pollution benefit function in Figure 1.1.

Fischer, S., R. Sahay and C. Vegh (1966), 'Stabilization and Growth in Transition Economies: The Early Experience', *Journal of Economic Perspectives,* **10** (2), Spring: 45–66.

Fournier, B. (1995), *Environmental Taxes in OECD Countries,* Paris: OECD.

Goldman, M. (1985), 'Economics of Environment and Renewable Resources in Socialist Systems Part I: Russia', in A. Kneese and J. Sweeney (eds), *Handbook of Natural Resource and Energy Economics.* Vol 1, Amsterdam: North-Holland, pp. 725–45.

Hahn, R.W. (1989), 'Economic Prescriptions for Environmental Problems: How the Patient Followed the Doctor's Orders', *Journal of Economic Perspectives,* **3**: 95–114.

Hahn, R.W. and G.L. Hester (1989), 'Marketable Permits Lessons for Theory and Practice', *Ecology Law Quarterly,* **16**: 361–406.

Hahn, R.W. and R.N. Stavins (1992), 'Economic Incentives for Environmental Protection: Integrating Theory and Practice', *American Economic Review,* **82**: 464–8.

Hertzman, C. (1993), 'Environmental Epidemiology in Central and Eastern Europe', *Health and Environment Digest,* **6** (10): 1–40.

Howe, C. (1994), 'Taxes versus Tradeable Discharge Permits: A Review in Light of the U.S. and European Experience', *Environmental and Resource Economics,* **4** (2): 151–70.

Hughes, G. (1992), 'Are the Costs of Cleaning Up Eastern Europe Exaggerated? Economic Reform and the Environment', *Oxford Review of Economic Policy,* **7**(4): 106–36.

Larson, B.A. and G.B. Frisvold (1996), 'Uncertainty over Future Environmental Taxes: Impact on Current Investment in Resource Conservation', *Environmental and Resource Economics,* **8**: 461–471.

Lehoczki, Z. and G.E. Morris (1994), 'Environmental Fund Disbursement: Incentives for Performance in Central and Eastern Europe', Mimeo.

Lehoczki, Z. and G. Peszko (1994), 'Environmental Funds in the Transition to a Market Economy', Paper presented at the OECD Conference on Environmental Funds, St. Petersburg, October.

Lehoczki, Z. and G. Peszko (1995), *The St Petersburg Guidelines on Environmental Funds in the Transition to a Market Economy,* Paris: OECD.

Lovei, M. (1994), 'Environmental Financing: The Experience of OECD Countries and Implications for Transition Economies', Mimeo, October.

Milliman, S.R. and R. Prince (1989), 'Firm Incentives to Promote Technological Change in Pollution Control', *Journal of Environmental Economics and Management,* **17**: 24–65.

Mills, E.S. and P.E. Graves (1986), *The Economics of Environmental Quality,* 2nd edn, New York: W.W. Norton & Company.

Opschoor, J.B. and H.B. Vos (1989), *Economic Instruments for Environmental Protection,* Paris: OECD.

Pezzey, J. (1988), 'Market Mechanisms of Pollution Control: "Polluter Pays", Economic and Practical Aspects', in R. Kerry Turner (ed.), *Sustainable Environmental Management: Principles and Practice,* Chapter 9, Boulder, CO: Westview Press, pp. 190–242.

Potier, M. (1995), 'The Experience of OECD Countries in their Domestic Use of Economic Instruments for Environmental Management,' in B. Moldan (ed.), *Economic Instruments for Sustainable Development,* Ministry of Environment of the Czech Republic.

Roberts, M.J. and M. Spence (1976), 'Effluent Charges and License Under Uncertainty', *Journal of Public Economics,* **5**, April/May: 193–208.

Sachs, J. (1995), 'Economies in Transition: Some Aspects of Environmental Policy', Environment Discussion Paper No. 1, Harvard Institute for International Development, February.

Slocock, B. (1992), *The East European Environment Crisis: Its Extent, Impact and Solutions,* Economist Intelligence Unit Special Report No. 2109, Business International, Ltd, London.

Smith, S. and T. Zylicz (1994), 'Rapporteurs' Report of the Workshop on Taxation and Environment in European Economies in Transition', in *Taxation and the Environment in European Economies in Transition,* OECD; pp. 6–15.

Tietenberg, T.H. (1980), 'Transferable Discharge Permits and the Control of Stationary Source Air Pollution: A Survey and Synthesis', *Land Economics,* **56**: 391–416.

Tietenberg, T.H. (1992), *Environmental and Natural Resource Economics,* 3rd edn, New York: Harper-Collins.

Toman, M. (1993), 'Using Economic Incentives to Reduce Air Pollution Emissions in Central and Eastern Europe: The Case of Poland', *Resources,* Resources for the Future, Fall.

Weitzman, Martin (1974), 'Prices vs. Quantities', *Review of Economic Studies,* **41**, October: 447–91.

World Bank (1996), *From Plan to Market: World Development Report 1996,* Oxford: Oxford University Press.

World Environment Center (1995), *Economic and Environment Benefits of Industrial Waste Minimization in Poland*, World Environment Center.

World Environment Center (1995), *Economic and Environment Benefits of Industrial Waste Minimization in Estonia, Lativa ans Lithuania*, World Environment Center.

Zylicz, T. and Z. Lehoczki (1993), 'Towards Environmental Recovery. The Czech Republic, Hungary, Poland, Slovakia', Mimeo, September.

APPENDIX 1A IMPACT OF POLLUTION CHARGES AND TECHNICAL CHANGE ON OPTIMAL POLLUTION LEVELS: OUTPUT-DEPENDENT AND INPUT-DEPENDENT POLLUTION MODELS

The purpose of this appendix is to show how pollution charges alter the marginal benefits of pollution to enterprises and to illustrate how investment in pollution control also affects marginal benefit functions. Two cases are considered, the output-dependent case and the input-dependent case. The models are developed in a very simple way to explain the basic issues, and it is assumed that the reader has a good knowledge of the duality between production functions and profit functions.

I Output-dependent Pollution

In this model y is output; k is a capital input with rental price r; l is labour input with wage rate w; $y = f(k, l)$ is the neoclassical production function, e is pollution, t is the unit pollution tax, and $e = (1/A)f(k, l)$. With this common formulation, pollution generation depends on output levels, with the parameter $1/A$ interpreted as the pollution intensity of output.

Assuming the enterprise maximizes profits, the enterprises problem can be written as:

$$\max_{k,l,e,y} py - rk - wl - te \text{ s.t. } y = f(k, l) \text{ and } e = (1/A)f(k, l). \tag{1A.1}$$

Substituting the constraints directly into the objective function, the enterprises problem can be written as:

$$\max_{k,l} (p - t/A)f(k, l) - rk - wl. \tag{1A.2}$$

From the firm's problem in (1A.2), it is clear that a pollution charge t acts exactly like a decrease in the output price, where the charge is weighted by the pollution intensity parameter $1/A$. The solution to problem (1A.2) implies a standard indirect profit function of the form:

$$\Pi = \Pi(p - t/A, w, k). \tag{1A.3}$$

To find the optimal level of pollution, the envelope theorem (Hotellings lemma) implies:

$$\frac{\partial \Pi}{\partial t} = -\frac{1}{A}\frac{\partial \Pi}{\partial p'} = -\frac{y^*(p',w,r)}{A} = -e^*(p',w,r) \qquad (1A.4)$$

where $p' = p - t/A$, and y^* and e^* are the optimal levels of output and pollution for the firm given the prices of outputs, inputs and the pollution charge. From (1A.4), this optimal pollution function e^* can also be interpreted as the marginal pollution abatement cost function.

For the output-dependent problem, there are two main questions of interest. First, what is the impact of a higher charge t on pollution? And second, what is the impact of improved pollution control, which reduces the pollution intensity of output (a higher A), on pollution?

To answer question 1, the optimal pollution function (1A.4) can be used to show that:

$$\frac{\partial e^*}{\partial t} = -\frac{1}{A^2}\frac{y^*}{p'} < 0 \text{ and } \frac{\partial y^*}{\partial t} = -\frac{1}{A}\frac{y^*}{p'} < 0. \qquad (1A.5)$$

So, from (1A.5) we know that a higher tax always reduces pollution and output.

Since information on the response of pollution to a higher charge is often of interest before a tax is chosen, (1A.5) can also be used to show that:

$$n_{e,t} = -\frac{t}{Ap} n_{y,p} < 0 \qquad (1A.6)$$

where the notation $n_{e,t}$ and $n_{y,p}$ are used to represent elasticities of e with respect to t and y with respect to output price p.

In equation (1A.6), the output elasticity is always positive (or zero), and the term $t/Ap = (t/A)/p$ is the ratio of the tax per unit of output relative to output price. In general t/Ap is between zero and one. As a result, the pollution demand elasticity can be expected to be inelastic (between –1 and zero) when supply is own-price inelastic. It is possible for the pollution elasticity to be less than –1 when supply is elastic and the term t/Ap is large enough.

What is important from (1A.6) is that simple information on the charge level, output price, pollution intensity and output supply response can be used to predict the pollution response to a changed charge level. For example, if $1/A = 0.5$ is kilograms of pollution per unit of output (pollution intensity), the tax is $2 per kilogram, the output price is $10 per unit of output, and the output supply elasticity is 2, then the pollution elasticity with respect to the charge is –0.20. In this example, the output elasticity would have to be 5 or larger for the pollution elasticity to be less than or equal to –1.

The second issue to consider is the impact on pollution levels of improved pollution control, which reduces the pollution intensity of output (a higher A). To answer this question, (1A.4) can be used to show that:

$$\frac{\partial e^*}{\partial A} = -\frac{y}{A^2}(1+n_{e,t}) \qquad (1A.7)$$

which can also be written as:

$$n_{e,A} = -(1+n_{e,t}). \qquad (1A.8)$$

It is also possible to substitute (1A.6) for the pollution elasticity in (1A.8) to write the equation in terms of basic economic variables that are more easily observable.

From (1A.8), if pollution does not respond to the charge, so that $n_{e,t}$ is zero, then a one per cent increase in A yields a one per cent decrease in pollution. In general, however, if there is some responsiveness of pollution to the charge, then a one per cent increase in A yields less than a one per cent decrease in pollution. In fact, it is possible that pollution levels will increase with an increase in A (decrease in pollution intensity) when $n_{e,t} < -1$. This is why in Figure 1.4 the MB(new') function is above MB(old) for some levels of pollution.

The analysis in (1A.1)–(1A.8) can be simulated with a CES production function of the form:

$$\dot{y} = f(k,l) = \left[k^P + l^P\right]^{h/\rho}. \qquad (1A.9)$$

To investigate question 1, the charge can, for example, vary between 0 and 70. Reasonable parameter values might be the following: $P = -2; h = 0.70; w = 2; r = 1; p = 100; A = 2$. For the CES production function, $h < 1$ implies decreasing returns to scale and $\rho = -2$ implies an elasticity of input substitution of $-1/3$, which is between Cobb–Douglas (-1) and Leontief production functions (0). Question 2 can also be investigated by evaluating changes in the pollution intensity for a one unit increase in A to a value of 3.

As a final point, which is especially relevant for some economies in transition, if some inputs are relatively fixed in the short run, the demand for pollution can become very 'steep' and less responsive to changes in pollution charges. This result can be shown by simulating the model holding the level of labour l fixed at the optimum when $t = 0$, and then gradually increasing t while holding l constant.

II Input-dependent Pollution Generation

For some types of pollution problems, it is probably better to think of pollution being generated by some specific input or set of inputs, which are then combined with other inputs to create output. Examples include CFCs, water and energy inputs.

Such a situation can be modelled as:

$$\max_{l,e} \; pf(l,be) - wl - (r+t)e \tag{1A.10}$$

where p is output price, f is the production function, l are inputs not related to pollution, e is pollution, b is a technological efficiency parameter, w is the input price, r is the aggregate input cost per unit of pollution, and t is a pollution charge. The technology parameter b translates e units of pollution into units of 'effective' pollution. Thus, it is possible to hold output constant when e decreases if b increases. While not necessary to develop for the purposes of this appendix, the unit input price r can be derived using index-theory, with r being the unit cost function from minimizing polluting input costs for one unit of pollution.

Using the same methods as followed for the output-dependent case, it can be shown that:

$$\frac{\partial e^*}{\partial t} < 0 \text{ and } n_{e,b} = -\left[1 + \frac{(r+t)}{t} n_{e,t}\right]. \tag{1A.11}$$

This input-dependent case can be simulated as well using the general CES formulation. It is very easy to simulate the case where increased efficiency in the use of pollution (that is, increases in b) leads to increased pollution in part because the term $(r + t)/t$ in (1A.11) is always greater than one.

2. A Survey of Pollution Charge Systems and Key Issues in Policy Design

Jeffrey Vincent and Scott Farrow

1 INTRODUCTION

Environmental charge systems have the unusual potential to achieve several objectives simultaneously. Such charges can reduce pollution to a defined target, minimize the cost to the economy of achieving that target (cost effectiveness), and raise money for the government. There is now considerable experience in transition economies in Europe and the former Soviet Union with trying to implement such charge systems on a wide variety of pollutants to meet these multiple objectives.

To a large degree, existing charge systems in the region have evolved from Soviet-era predecessors. It is now well known that environmental policy in the Soviet system often was strong in law but was not implemented, enforced or achieved (see, for example, Ahlander, 1994; National Academy of Public Administration, 1994; Danilyan et al., 1993). As a better understanding of environmental problems and, perhaps, the severity of such problems grew within the Soviet Union, understanding also grew of alternative approaches for setting and attaining environmental goals (see National Academy of Public Administration, 1994). Charge systems began to be discussed in the 1970s, research on environmental costs and benefits occurred in the early 1980s, and these developments joined with greater public interest in environmental issues in the mid-1980s.

While some regions within the Soviet Union had instituted some form of pollution charges earlier, in 1988 the Soviet government issued a decree identifying pollution charges as part of a radical change that included the creation of a new, top-level, environment committee, *Goskompriroda*. Draft regulations were developed for pollution charges in 1989, and pilot projects were introduced in 49 regions in 1990. Pollution charges were introduced throughout the Russian Federation in early 1991. The development and implementation of charge systems has therefore been part of an evolutionary process in environmental policy design that began decades ago and continues today. This process also occurred in centrally-planned economies in Europe during the same period.

The purpose of this chapter is to provide an overview of pollution charge systems that now exist in transition countries of Europe and the former Soviet Union. It draws on information from eleven countries: Belarus, Bulgaria, the Czech Republic, Estonia, Kazakhstan, Latvia, Lithuania, Hungary, Poland, Russia and Slovakia. Section 2 summarizes the key features of the systems, including the structure of charges, the reasons for having charges, the types of pollutants covered and how the charge revenues are used. Some information is also provided on how charge rates are determined, how they are adjusted for inflation and how they are enforced. Section 3 then discusses three sets of issues regarding the design of charge systems: their cost effectiveness; their impacts on enterprises' financial viability; and their impacts on productive investments. Section 4 ends with a brief discussion of the opportunity to integrate charge systems with tradable permit systems.

2   KEY FEATURES OF POLLUTION CHARGE SYSTEMS

2.1   Basic Features

Charges are levied in the eleven countries on both air and water pollutants. They are linked to permissible limits, which are specified in pollution permits and are defined in terms of quantity of pollution per time period (for example, tons per year or quarter). Charge rates are based on similar units. Charges usually include both a base rate for emissions within the permissible limits and a higher rate – sometimes called a 'penalty' rate – for emissions above the limits. That is, they are two-tiered.

Permissible limits are best viewed as simply specifying the level at which the charge rate increases. Enterprises are legally allowed to pollute above the limits, as long as they pay the higher rate associated with such levels. Thus, it is misleading to refer to penalty rates as 'fines', 'sanctions', or other terms that denote a legal violation. In some countries, the base rate is zero and only a penalty rate applies. We refer to these simpler systems as noncompliance fee systems.

Charge rates may be either uniform or progressive. With uniform systems, a constant rate of t is paid for emissions within permissible limits, and a constant rate of m is paid for emissions above limits, so that $m = kt$, with k being the 'penalty' multiple. In noncompliance fee systems, $t = 0$, and m is set at a positive level. When the enterprise is within its permissible limit, Q', the total payment is $t*Q$, where $Q \leq Q'$ and Q is the level of actual emissions. When the enterprise is above its limit, the total payment is usually $tQ' + m(Q - Q')$. In some countries, however, enterprises pay mQ when above their limits (that is, the penalty rate is assessed on all units, not just the units above the limit). With progressive systems, the rate depends on the level of pollution, so that $t = t(Q)$ and $m = m(Q)$, where the functions $t(Q)$ and $m(Q)$ determine how the rates vary with pollution levels. These functions are increasing in Q. Variations of these systems are explained more fully in the individual country chapters.

Charge rates are usually determined at the national level, while pollution permits and the limits in those permits are usually determined and implemented by regional environmental authorities. In some cases a national agency such as the Ministry of Environment issues permits to the largest polluters. In principle, permissible limits are determined by working backwards from a desired level of ambient environmental quality. Such ambient quality goals may be formalized through national ambient standards, or they may be based on the opinions of the permitting authorities when such standards do not exist, are not legally binding, or for some other reason are ignored. In practice, permissible limits are often negotiated with enterprises, and they may vary within a country depending on ambient conditions and a variety of other reasons. Even in countries with uniform emission concentration standards across enterprises, permissible limits vary, and they vary widely. Continuous monitoring is rare, so the pollution load is typically calculated by multiplying estimated volumes of wastewater or stack gases times the estimated pollution concentrations. Volume and concentration estimates are sometimes based on point measurements, and sometimes by engineering analyses of inputs and outputs.

There is a close link between the two-tiered system and accounting procedures in former centrally-planned economies. In those economies, base rates were regarded as an ordinary operating cost for which enterprises could request reimbursement from the central authorities. Payments on pollution above the limits, however, had to be paid out of an enterprise's 'surplus' (value of output minus reimbursable costs). As long as enterprises were furnished funds to cover charges on pollution within the limits, only the penalty rate could create an incentive for pollution reduction. The reimbursement amounted to a pollution subsidization policy. With the collapse of central planning, base rates can now also provide some incentive for pollution control, depending upon their levels relative to abatement costs, other prices and subsidies, and the effectiveness of enforcement.

2.2   Charges on Air Pollution

Table 2.1 summarizes basic information on charge systems provided by country experts involved in developing and implementing pollution charge systems in the eleven countries.[1] Current charge systems for air pollution date from the early 1990s in most countries. In several cases, current systems have evolved from ones originally introduced decades earlier (for example, in the 1960s in the Czech Republic and Slovakia, and in the 1970s in Bulgaria, Hungary and Poland). The two-tiered system is the most common one in the region, occurring in nine of the eleven countries. The two exceptions are Bulgaria, which has a uniform noncompliance fee system, and Hungary, which has a progressive noncompliance fee system. Countries with the two-tiered system are about evenly split between ones with uniform charges and ones with progressive charges. In Lithuania and Slovakia, all charges are progressive.

Several objectives are often stated for the countries' charge systems. They include: (a) inducing compliance with permissible limits, (b) forcing polluters to pay for the use of the environment, and (c) raising government revenues, which are usually channelled back into some form of environmental expenditures. Not surprisingly, inducing compliance with limits is considered to be the most important objective in Bulgaria and Hungary, but this objective is also important in countries with two-tiered systems (for example, Latvia, Poland and Slovakia). In countries with two-tiered systems, penalty rates are 1.5 to 500 times base rates, with a factor of 10 being the median ratio.

Forcing polluters to pay for the use of the environment (the 'polluter pays principle') and raising revenues for environmental expenditures are also key reasons for charges in some countries. They are given as the most important reason in Belarus, the Czech Republic, Estonia, Kazakhstan and Russia. These two goals are tightly linked in the region, because the vast majority of charges are allocated to off-budget funds earmarked for environmental expenditures. In most countries, all charge revenue is allocated to national or subnational (regional or municipal) environmental

1   Much of this information was gathered at the Vilnius Workshop on Implementation of Pollution Charge Systems in Transition Economies, held 18–20 September 1995. The workshop was co-sponsored and organized by the Ministry of Environmental Protection, Republic of Lithuania, the Ministry of Environmental Protection and Regional Development, Republic of Latvia, and the Harvard Institute for International Development. Funding was provided by the United States Agency for International Development.

funds. Subnational funds receive the largest share of revenues in Belarus, Kazakhstan, Lithuania, Poland and Russia.

One can therefore sum up the overwhelming motivation for charge systems as the joint – but not totally consistent – desire for less pollution and more revenue. Cost effectiveness, the most important textbook objective, has not been an important motivation. It is mentioned as a motivation in some countries (for example, Lithuania), but such statements do not appear very convincing.

Charges are levied on a large number of pollutants, more than a hundred in seven of the countries. Charges on SO_2 (sulphur dioxide) generated the greatest share of revenue in most countries in 1994, followed by charges on particulates/dust. Rates in most countries must ultimately be approved by the highest levels of government, either the legislature, the cabinet, or a similar high-level interministerial body. In most countries they are proposed by the Ministry of Environment, which consults with various ministries in formulating them. The Ministry of Environment has substantial autonomy in setting rates only in Hungary and Lithuania.

While economics textbooks suggest that pollution charges should equate marginal environmental benefits and marginal abatement costs, few countries explicitly consider the economic value of damage to the environment or human health in setting charge levels. Some countries do, however, informally consider the value of environmental damages. For example, in Estonia, Hungary, Kazakhstan and Russia, rates vary within the countries according to either the level of existing pollution in a particular location or the sensitivity of the local environment to pollution, or both. This flexibility enables charge systems in the four countries to reflect local differences in environmental damage. In countries without this rate flexibility, adjustments for differences in environmental damage can be made only through differences in permissible limits.

More emphasis is placed on the financial impacts of charges on enterprises (compliance costs). Consistent with the importance of revenue as a motivation for the systems, Ministries of Environment try to keep charges high enough to cover a meaningful portion of the costs of government environmental programmes, but they face pressure to keep charges low to minimize the financial impacts on enterprises. Despite the stated concern about these impacts, however, governments often do not consult directly with enterprises when rates are determined. Moreover, only a few studies on pollution-abatement costs or the ability of enterprises to pay pollution charges have been conducted (for example, in the Czech Republic, Poland, Estonia and Russia.) Hence, rates are generally set in a situation where the environmental benefit side is largely ignored and the abatement cost side is poorly understood.

Table 2.1   General features of charge systems for air and water pollution in 1994

	Medium	System	Main objective	Penalty multiple	Revenue allocation[a]
Belarus	Air	TT–U	PPP	15	SF–90%
	Water	TT–U	PPP	15	NF–10%
Bulgaria	Air	NC–U	C	n.a.	NF–70%
	Water	NC–U	C	n.a.	SF–30%
Czech	Air	TT–U	R	15	NF–100%
NF	Water	TT–P	C	5	
Estonia	Air	TT–P	PPP	5–500	NF–50%
NF	Water	TT–P	C	5–1,000	SF–50%
Hungary	Air	NC–P	C	n.a.	NF–70%
	Water	NC–P	C	n.a.	SF–30%
Kazakhstan	Air	TT–P	PPP	1–10	SF–65%
	Water	TT–P	PPP	1–10	NB–20%
					NF–15%
Latvia[b]	Air	TT–U	C	4	NB–30%
	Water	TT–U	C	4	SB–70%
Lithuania	Air	TT–P	CE	10	SF–70%
	Water	TT–P	CE	10	NB–30%
Poland	Air	TT–U	C	10	SF–64%
	Water	TT–P	C	4[f]	NF–36%
Russia	Air	TT–U	R	5	SF–54%
	Water	TT–U	R	5	NF–36%
					NB–10%
Slovakia	Air	TT–P	C	1.5	NF–100%
	Water	TT–P	C,R	3	

Notes
TT–U   Uniform two-tiered system
TT–P   Progressive two-tiered system
NC–U   Uniform noncompliance fee system
NC–P   Progressive noncompliance fee system
C   Compliance
CE   Cost effectiveness
PPP   Polluter pays principle
R   Revenue
NB   Central government budget
NF   National environmental fund
SB   Subnational (provincial or municipal) budget
SF   Subnational (for example, provincial or municipal) environmental fund
CC   Compliance costs
D   Environmental damage

No. of pollutants charged	Revenue leaders	Charges vary/ w/in country?	Factors influencing charge levels	Inflation indexed
>150	No Data	No	CC,D,PR,R	Yes
>150	No Data	No	CC,D,PR,R	Yes
16	SO$_2$	No	CC	No
27	BOD/COD	No	CC	No
90	SO$_2$	No	CC,D,PR,R	No
5	BOD	No	CC,D,PR,R	No
139	SO$_2$	Yes	CC,R	Yes
8	BOD	Yes	CC,R	Yes
150	SO$_2$	Yes	CC,D,PR,R	No
32	No Data	Yes	R	No
>100	No Data	Yes	CC,PR,R	Yes
>100	No Data	Yes	CC,PR,R	Yes
7	SO$_2$	No	D,PF	No
10	BOD	No	D,PF	No
All	No Data	No	CC,D	Yes
All	No Data	No	CC,D	Yes
62[c]	SO$_2$	No[d]	D,R	Yes[e]
6	BOD/COD	Yes	D,R	Yes
>100	No Data	Yes	CC,PR,R	Yes
>100	No Data	Yes	CC,PR,R	No
123	Particulates	No	CC,PR,R	No
5	BOD	No	CC,PR,R	No

PF   Political factors
PR   Enterprise profitability
R   Government revenue needs
a.   These are the basic revenue allocations, but there are certain differences for specific pollutants in some countries. Figures apply to revenue from both water and air charges.
b.   Seventeen pollutants are specified in the Regulations of the Cabinet of Ministers, drafted for the Law On Natural Resources Tax.
c.   Plus seven sources of evaporate emissions associated with refilling fuel storage tanks.
d.   In the early 1990s, charges were twice as high in Katowice and Krakow, to take into account the local level of pollution and to raise additional revenue for regional environmental funds.
e.   In October–December, projections of inflation during the upcoming year are used to revise charges.
f.   Approximate. Not a simple multiple.

All countries except Bulgaria rely partially or completely on enterprises to report the amounts of pollution they discharge, with regional environmental inspectorates or some other unit of the Ministry of Environment responsible for monitoring and enforcing the self-reporting system. Revenue (tax) services also play an enforcement role in Belarus, Kazakhstan and Latvia. Government agencies responsible for enforcement can levy fines on enterprises that fail to pay their charges. In some countries, agencies can also withdraw funds from the enterprises' bank accounts (Belarus, Bulgaria, Kazakhstan, Russia), take the enterprises to court (Latvia, Estonia, Slovakia, Hungary), or, in extreme cases, shut them down (Czech Republic, Russia).

Despite the existence of charges and stern measures for enforcing them, country experts agree that charges have not created a strong incentive for pollution-control investments. There are a variety of reasons for this. These reasons are discussed in more detail in the country-specific chapters, but two are worth noting here. First, aside from Poland, the countries have very low charge rates, especially for pollution within permissible limits (and Bulgaria and Hungary have only penalty rates). This problem is exacerbated in some cases by permissible limits being so high that enterprises can remain within the limits even if they undertake no pollution control. Moreover, the real value of base rates (and penalty rates too) is reduced over time by inflation in the many countries without inflation-adjustment mechanisms (see Table 2.1). In countries without inflation-adjustment mechanisms, the date when charges were last revised varies greatly. In the case of Hungary, charges were last revised a decade ago. Environmental experts in the region may feel strongly that base rates are below marginal environmental damage costs, but their concerns are outweighed in the rate-setting process by concern over the financial impacts of charges on enterprises.

Second, monitoring systems are rudimentary, and enforcement is weak. Self-reporting breaks down because incentives for under-reporting clearly exist and environmental authorities do not have the resources to conduct regular, unannounced checks. Environmental authorities are among the weakest government agencies in transition economies, and this inhibits their ability to take action against violators, especially when the violators are state-owned enterprises.

2.3 Charges on Water Pollution

Charge systems for water pollution are quite similar to those for air. Five differences are worth noting. First, more countries have two-tiered systems with progressive charges. Second, compliance with permissible limits is stated somewhat more often as an important reason for the introduction of charges. Third, charges tend to be levied on fewer pollutants, although the absolute number is still high. Fourth, Poland is added to the list of countries with charges that vary within the country. Fifth, the charge systems are considered to be somewhat more effective, which probably reflects both higher charges and the greater ease in monitoring discharges of common pollutants.

3   ISSUES IN POLICY DESIGN

This section focuses on three basic economic policy issues: Sections 3.1 and 3.2 discuss the short-run issues of cost effectiveness and financial impacts, respectively, while Section 3.3 discusses the longer-run issue of investment efficiency. Issues of uncertainty regarding the value of damages caused by pollution and the costs of pollution control are not explicitly treated.

Two preliminary points are worth emphasizing regarding marginal pollution-abatement costs. First, as discussed in Chapter 1, the marginal cost of abating pollution is identical to an enterprise's demand for emitting pollution, which is affected by the same factors as any other input demand. Hence, abatement costs respond to changes in underlying prices and quantities, parameters and relationships; they are not fixed and constant. Second, if enterprises are not motivated by profit maximization, or at least by minimizing the cost of producing some target level of output, then marginal pollution-abatement costs are likely to be higher than necessary. Privatization, or at least hard-budget constraints and the right to retain earnings if enterprises are government owned, is the best way to strengthen the profit motive. The practical importance of this point is indicated by the fact that industrial enterprises are still state owned in most transition economics. Moreover, in some countries the bulk of pollution charges are levied on suppliers of public services, such as district heating, hot water and electricity, which are seldom private enterprises. Hence, there is reason to believe that abatement costs in the region are unnecessarily high, reducing the incentive effect of charges and increasing the financial difficulty of achieving permissible limits.

3.1   Cost Effectiveness

In a textbook model of pollution control (see Chapter 1), a single, uniform pollution charge levied on all polluters is a cost-effective approach to pollution control: it achieves a given level of pollution abatement at the lowest possible total cost of pollution control across all polluters. Cost effectiveness is especially important in transition economies, where financial resources are scarce and enterprises have opportunities to make productive investments that could generate additional income and employment, which are major concerns of transition-country citizens and governments.

The first question to ask regarding the cost effectiveness of charge systems is whether all polluters are paying the same charge on their last unit of pollution, which is the theoretical condition required for cost effectiveness. Given the widespread use of two-tiered systems, the answer is obviously no. Polluters below or at permissible limits have lower abatement costs at the margin, equal to the base rate, while those above their limits have higher marginal costs, equal to the penalty rate. Even if all polluters face the same permissible limit, their abatement cost functions can still vary, potentially causing some to be above the limit and some to be below and creating differences in marginal abatement costs.

In practice, however, several factors can reduce the magnitude of the loss in cost effectiveness. First, in some countries, permissible limits for certain pollutants may be set so high that all or nearly all polluters are within them and are paying the same base rate at the margin. Estonia and Latvia provide two examples where this is the case. In such cases, the two-tiered system is effectively single tiered, and it is cost effective as long as no enterprises emit beyond their limits.

Second, and at the opposite extreme, the base rates in some countries are so low that even the tenfold median difference between them and the penalty rates is not very large in absolute terms. For instance, while biological oxygen demand (BOD) base rates are about $700 per ton in Poland, in Latvia they are on the order of just $56 per ton. Base rates for sulphur dioxide and nitrogen oxides are only slightly above $1 per ton in Russia. Under these circumstances, the two-tiered system again effectively collapses to a single-tiered charge, only now it equals the penalty rate, with few firms complying with their permissible limits. The loss in cost effectiveness, then, depends on whether other measures are being used to force compliance, and how these measures vary across polluters. If these measures lead to more control than would result from the low charges alone, then cost effectiveness may be sacrificed.

Third, negative impacts on cost effectiveness may be mitigated in countries that establish permissible limits by negotiating with enterprises. If high-cost abaters are granted a somewhat higher limit than low-cost abaters, then total pollution-abatement costs will be lower than if the limits were identical. Of course, negotiations result in a more cost-effective allocation of abatement responsibility only if the basis of the negotiations is the polluter's abatement cost. Although these costs are indeed considered when determining permissible limits in the region, they are not the only factors considered; political considerations enter, too.

3.2 Financial Impacts and Political Acceptability

Raising charge rates to levels high enough to achieve environment ministries' pollution control objectives would have financial impacts on polluters that are politically unacceptable in most countries. One way to reduce the negative financial impacts of higher rates would be to modify the two-tiered charge system so that it provides the correct incentives at the margin but reduces the total payments made by the enterprise. This can be implemented by providing a 'charge credit' or a fixed reduction in charge payments.

Farrow (1995a) has shown generally, and Pezzey (1988) has demonstrated for a particular case, that the financial burden can be reduced by granting a charge credit in the amount of AtQ^0. As before, t is the charge rate; A ($0 \leq A \leq 1$) is a politically chosen revenue parameter, and Q^0 is a fixed quantity of emissions, perhaps equal to the permissible limit.[2] The total charges to be paid would be $tQ - AtQ^0$, which can be negative, that is, a subsidy, if Q is sufficiently small. The correct incentive for cost effectiveness is provided because each additional unit of emissions costs the enterprise the full amount t. Although superficially similar to the two-tiered system in that it appears that a lower charge is being paid for emissions up to Q^0, the incentive is actually that of a uniform charge because the credit is granted no matter what quantity of pollution is actually emitted. In effect, the government is collecting an amount t per unit of pollution discharged above the fixed level AQ^0 and providing a payment of amount t per unit abated below AQ^0.

A less formal charge–credit system, which does not have cost-effective properties, exists in the Czech Republic, Estonia, Latvia and Lithuania: enterprises are allowed to retain some of their pollution charge payments to finance approved pollution-control investments. This system is not cost effective because it is simply added to the existing two-tiered system, which as noted in the previous section is inherently not cost effective. It does not eliminate the financial burden on enterprises either, because an enterprise must either pay the charge or make some approved type of investment. It does, however, alter the enterprise's cash-flow situation, making it less reliant on external financial resources for financing pollution-control in-

---

2   As discussed in Chapter 1, Pezzey (1988) suggests setting the parameters so that the total tax payment equals the total value of environmental damages at the economically optimum level.

vestments. This system requires that environmental authorities determine the types of investments that qualify for a credit, monitor whether investments actually occur, and establish a credible enforcement mechanism if the credit is not used appropriately. Of particular importance is whether credits should be approved only for end-of-pipe investments or also for 'productive' investments that have a positive environmental impact. When environmental authorities have the main approval authority for granting credits, it is likely that end-of-pipe investments will be preferred, because such technologies are familiar to the authorities and are relatively easy to observe and verify. However, there are many situations where a modest investment intended to improve production efficiency also has a positive environmental impact. If totally free to allocate investment resources, an enterprise would be expected to prefer productivity-enhancing investments. Charge–credit systems that do not allow this option might cause enterprises to invest too much in end-of-pipe approaches and thereby increase the cost of pollution control.

3.3 Investment Efficiency

Although the impacts of environmental policies on investment behaviour have been analysed in the environmental economics literature,[3] the analyses often focus narrowly on pollution-abatement investments. In transition economies, where much of the productive capital stock is greatly outdated, one cannot ignore the issue of the impacts of charges on productive investments. In addition, a substantial amount of environmental investment is made by state environment funds, which may not target the investment opportunities that earn the highest returns.

These issues are addressed more easily in theory than in practice. While this is not the place to develop such an analysis, theoretical considerations suggest the following conclusions.

1. Even after accounting for the opportunity cost of alternative, productive investments, a pollution charge is necessary for striking the optimal balance between production of goods (which generates pollution) and current benefits from environmental protection.
2. When opportunities exist for private pollution-abatement investments or productive investments that improve environmental quality, a polluting industry should be allowed to credit against its total pollution charge some portion of its expenditure on such investments. (Alternatively, an environmental fund that returns some

3   See, for example, Milliman and Prince.

portion of charge revenue to industry should be established.) The magnitude of the credit depends on the relative returns to public and private environmental investments. The optimal credit increases as the returns to private environmental investments rise relative to the returns to public environmental investments.

3. The last point implies that governments should retain funds for public environmental investments only if those investments yield a return greater than the return to private environmental investments.

4. The levels of the pollution charge and the credit for private investments depend on the returns to productive investments. As those returns rise: (i) the charge should be smaller (industry should pay less in the first place), and (ii) the credit should be larger (industry should retain more of the charge, to liberate earnings for productive investments).

These findings are not surprising. Nevertheless, they point to three issues that merit greater attention in transition economies: (i) whether governments and environmental funds indeed invest the revenue they receive from charges more efficiently than industry would, (ii) whether industry uses funds from charge credits, in countries where they exist, for investments that potentially can yield environmental improvements, and (iii) whether the industry's investments are effective in practice (for example, industry might invest in abatement technology but not actually operate the equipment).

4   POLLUTION CHARGES AND TRADABLE PERMITS

While there are several variations of pollution charge systems implemented in the countries discussed in this chapter, each of those systems is essentially an add-on to a system of pollution permits that had already existed for some years. The permits specify a performance standard of some kind (the permissible pollution limit), but they do not specify specific technologies to achieve that standard. Thus, at least to some extent they allow cost considerations to enter into the choice of abatement strategy. Ironically, this is more flexible than the regulatory approach in many OECD countries. Furthermore, the existence of the permitting system suggests an alternative to efforts to improve the charge system: to introduce further flexibility by making the permits tradable.

It is an easy and logical generalization to move from permits issued on a source-by-source basis (for example, each stack at a factory) to enterprise-level permits, essentially an enterprise-level bubble policy, so that enterprises have increased flexibility to adjust production methods to achieve the same level of pollution at the site for less cost. Estimates of cost savings in the United States suggest that this kind of flexibility is the largest source of potential cost savings. This flexibility already exists in some countries (for example, air permits in Lithuania and Estonia). Thus, some amount of informal intraenterprise pollution trading already occurs in the region. This kind of trading could be expanded significantly.

A further modification of existing permits would be to develop permitting processes that allow pollution trades across enterprises.[4] Such trading would involve two or more enterprises that are jointly owned by the same entity (including the state) or independently owned. Under a simple trading system, enterprises can meet their emission targets by accomplishing on-site reductions (via process changes, retrofits, curtailments or shutdowns) or by paying other enterprises or small sources to reduce their emissions. An example of a variation on this system is to offer enterprises delayed compliance in return for funding conversion of household heating systems from coal-burning to less-polluting energy sources (Farrow and Bluffstone, 1995). The additional flexibility provided by a tradable permit programme could assist enterprises in making a transition to new technologies and obviate the need to install expensive pollution-control equipment on machinery that is planned for replacement.

Like a well-designed charge system, tradable permit systems promote cost effectiveness, but they are not a costless way of improving the environment. Some enterprises still have to invest in pollution control, and the government still needs to enforce sanctions against enterprises that discharge above the level of permits they hold. A trading programme should not be implemented if industry support is based on the idea that nobody will have to pay or if enforcement cannot ensure a reasonable level of compliance. As permits are typically granted free of charge instead of sold, enterprises initially must pay only for control costs, and therefore they might prefer trading to some types of charge systems. Once trades occur, expenditure on pollution permits from trading stays within industry instead of going to the government.

Numerous details must be in place before existing permit systems can be modified to allow interenterprise trading. Examples of these details are highlighted in Table 2.2, which comes from Margolis et al. (1995), based on a preliminary design for an air emission trading programme in Almaty, Kazakhstan. This programme became operational

---

4   There is a large literature on pollution trading. See, for example, Tietenberg (1985), Hahn (1989), Atkinson and Tietenberg (1991), Burtraw (1995) and Farrow (1995a,b,c).

Table 2.2   Design elements: draft emissions trading programme, Almaty, Kazakhstan

Design element	Almaty's preliminary choice
1. Goal of programme	Reduce SO_x, SPM, and formaldehyde from permitted sources by 7 per cent per year
2. Geographic region	Entire city with hot-spot provisions
3. Emission sources included	1,200 existing industrial and heat sources and new sources with same criteria
4. Emissions baseline or cap	41,000 tons (based on 1994 and 1991 emissions)
5. Allocation of allowances	Proportional share of baseline
6. Rate of emission reductions	Consistent with goal (7 per cent reduction)
7. End point of programme	Not defined
8. Nature of air credits	Allowances issued on a seasonal basis
9. Who is eligible to trade	Anybody, any enterprise inlcuding existing permitted sources who may buy and sell
10. Role of mobile source credits	Mobile source-derived credits may be used by industrial and heating sources
11. Trading process	Trades privately arranged, city may register liens
12. Reconciliation	Allow for 30 days after shortfall discovered, nominal fines resolution
13. Role of emissions banking	Allowed
14. Emission quantification protocols	Required but not yet defined
15. Monitoring, recordkeeping and reporting	Required but not yet defined
16. Confiscation protection	Allowances protected; but can be restricted
17. Lifetime of air credits	3 years
18. Enforcement provisions	Seller liable if allowances are double used
19. Programme financing	Self-financed via allowance/trade fee
20. Programme administration	Administered by city
21. Audits and backstops	Not yet defined
22. Programme phase-in	Phased in as required elements put in place

Source: Margolis, Trivedi and Farrow (1995).

in the winter of 1996. Interenterprise trading has also been explored in Poland.

Proposals also exist that recommend retaining a two-tiered charge system in combination with enterprise emission limits that can be met by trading (Roberts and Spence, 1976). Such a system would preserve many of the characteristics of the existing system in that both charges and permissible limits would be central elements of the programme. In such systems, the lower and upper tiers set the boundaries of the prices at which permits trade. A charge-and-trade system is more complex than a single system, but given the substantial uncertainty that exists in transition economies regarding abatement costs and environmental damages, it might be preferable to either a single-tiered charge system or a trading system. Existing regulatory systems in the region have many of the necessary components of such systems. As discussed in the Latvia chapter (Chapter 4), a new law in that country creates the opportunity to develop this type of combined system under certain circumstances.

REFERENCES

Ahlander, Ann-Marie Satre (1994), *Environmental Problems in the Shortage Economy: The Legacy of Soviet Environmental Policy,* Aldershot: Edward Elgar.

Atkinson, S and T. Tietenberg (1991), 'Market Failure in Incentive-Based Regulation: The Case of Emissions Trading', *Journal of Environmental Economics and Management,* **21**: 17–31.

Burtraw, D. (1995), 'Cost Savings Sans Allowance Trades? Evaluating the SO_2 Emissions Trading Program to Date', Resources for the Future Discussion Paper 95–30, Washington, DC.

Danilyan, V.I., Y.M. Arskii, V. Gejdos and Z. Stepanek (1993), *Economics and the Environment in the Former Soviet Union and Czechoslovakia,* Economic Commission for Europe, Discussion Paper, Vol. 2, No. 4, United Nations, New York.

Farrow, S. (1995a), 'The Dual Political Economy of Taxes and Tradable Permits', *Economics Letters,* **49**: 217–20.

Farrow, S. (1995b), 'The Czech Republic: Potential Cost Savings from Using Economic Instruments', Environment Discussion Paper 4, Harvard Institute for International Development, Cambridge, MA, March.

Farrow, S. (1995c), 'Comparing Economic and Administrative Methods: The U.S. Experience', in B. Moldan (ed.), *Economic Instruments for Sustainable Development,* Proceedings of a Workshop held at Pruhonice, Ministry of the Environment of the Czech Republic, pp. 68–84.

Farrow, S. and R. Bluffstone (1995), 'Implementable Options for Cost Effective Air Pollution Reduction in Northern Bohemia', Harvard Institute for International Development, Cambridge, MA, January.

Hahn, R. (1989), 'Economic Prescriptions for Environmental Problems', *Journal of Economic Perspectives,* **3** (2): 95–114.

Margolis, J., G. Trivedi and S. Farrow (1995), 'Feasibility Assessment for an Areawide Emissions Trading Bubble in the City of Almaty, Kazakhstan', Prepared by Dames & Moore for the Harvard Institute for International Development, Cambridge, MA, July.

Milliman, S.R. and R. Prince (1989), 'Firm Incentives to Promote Technological Change in Pollution Control', *Journal of Environmental Economics and Management*, **17**: 247–265.

National Academy of Public Administration (1994), *The Environment Goes to Market: The Implementation of Economic Incentives for Pollution Control*, Washington, DC: Academy Press.

Pezzey, John (1988), 'Market Mechanisms of Pollution Control: "Polluter Pays", Economic and Practical Aspects', Chapter 9 in R. Kerry Turner (ed.), *Sustainable Environmental Management: Principles and Practice*, Boulder, CO: Westview Press, pp. 190–242.

Roberts, M. and M. Spence (1976), 'Effluent Charges and Licenses Under Uncertainty', *Journal of Public Economics*, April/May: 193–208.

Tietenberg, T. (1985), *Emissions Trading*, Washington, DC: Resources for the Future.

3. Estonia's Mixed System of Pollution Permits, Standards and Charges

**Ljuba Gornaja, Eva Kraav,
Bruce A. Larson and Kalle Türk**

1 INTRODUCTION

Estonia has made a rapid transition to a market economy. As part of the fundamental realignment of relative prices and an open trade policy, industrial and agricultural output have fallen, energy use has fallen and energy prices have risen, and trade has been diversified away from the former Soviet states and towards the West. GNP began to grow again in 1994–95, however, indicating that the bottom of the transition period may be over.

This transformation has had a profound impact on point and nonpoint sources of water, air and solid-waste pollution. In agriculture, with the elimination of subsidies and price supports, the livestock herd, cultivated area and mineral fertilizer use have declined, which has reduced point and nonpoint sources of water pollution from the agricultural sector. The reduction in electricity use and production has substantially reduced emissions from the country's two main power plants, Baltic and Estonian, which are also two of the biggest polluters in the country.[1] For these two power plants, sulphur dioxide emissions have fallen from about 180,000 tons in the late 1980s to 149,000 in 1991 to 97,800 tons in 1994. Water consumption and effluent discharges from industry and agriculture have fallen as well, while household water consumption and wastewater dis-

---

1 Electrical energy production fell from 14.6 TWh in 1991 to 9.2 TWh in 1994. Electricity consumption fell from 7.0 TWh in 1991 to 5.2 TWh in 1994 (Statistical Office of Estonia, 1995). Mining of oil shale, the main fuel source for electricity generation, fell from 19.6 million tons in 1991 to 14.5 million tons in 1994 (Estonian Environmental Information Centre, 1991, 1993, 1994). Consumption of other energy sources, such as peat, heavy fuel oil and natural gas also fell substantially during the 1991–93 period.

charges have remained relatively unchanged (Estonian Environmental Information Centre, 1991, 1994).

While the transformation to a market economy has yielded some important environmental benefits, a revised system of environmental policy continues to be developed and implemented as one component of the overall transition to a modern market economy. For example, during 1993 alone, more than 85 formal policy statements targeted directly at environmental issues were passed by Estonia, including laws, regulations and rulings of the government, and regulations and decrees of the Ministry of Environment. Several other laws and regulations, such as income and value-added tax laws, were also passed which more directly influence economic activity and, as a result, the use of natural resources and the generation of pollution.

The goal of developing such policies now is to create the conditions so that environmental concerns can be correctly internalized into the decision-making process of new private enterprises, public-owned state and municipal enterprises undergoing privatization, and remaining public-owned enterprises. To achieve this goal, such policies need to be reasonable, credible and enforceable. It is well known that laws during Soviet times did not meet these requirements; laws were stringent on paper but not enforced. In fact, there was a confidential register of towns where ambient air quality violated ambient standards by at least tenfold. In 1988, this register included 103 towns (Timber Industry, 1990, p. 17). Care is needed to ensure that a similar situation does not arise with new environmental legislation in Estonia.

Major changes to Estonian pollution control policy took place in December 1993, 1994 and 1995 when the basic framework law for pollution charges and several regulations were revised and revised again. This system is best described as a mixed system, which is based on a combination of pollution permits, threshold levels, differential charge rates, self-reporting, some ambient and effluent standards, and penalties for noncompliance. There has been little systematic economic analysis of the pollution charge system in Estonia, either of the general implications of the law or of implementation experience and results.[2]

The purpose of this chapter is to provide an overview of the mixed system of permits and pollution charges in Estonia and to evaluate recent implementation experience focusing on air and water pollution.[3]

---

2   One exception is Ensmann et al. (1995). This current chapter is a substantially revised, updated and expanded version of the earlier report.

3   Pollution charges complement the system of administrative regulations. These regulations include pollution permits, maximum permissible limits, environmental quality stan-

Section 2 provides a general policy overview as well as some basic comments on the economic logic of the system and links to the general Estonian tax system. Section 3 focuses on air emissions and charges, including a synthesis of the basic permit decision process and its links to ambient air quality standards. Implementation experience related to air emissions emphasizes a specific feature of Estonia; namely that a very small quantity of large polluters concentrated in the northeast of the country account for the vast majority of point-source air pollution and paid pollution charges. Implications for the permit and charge system are discussed. Section 4 focuses on water effluent permits and charges, which includes the sequential revision of effluent standards and pollution charges during 1994 and 1995. The latest available data on the results of these changes are presented. Section 5 concludes with several recommendations for improving upon the current system of permits and charges, emphasizing issues of implementation, monitoring and enforcement.

2   ESTONIAN POLICY OVERVIEW

2.1   The Act on Pollution Charges

The basic framework law for pollution control is the Act on Pollution Charges (Resolution No. 244 of the President of the Republic of Estonia, 19 December 1993, effective 1 January 1994.)[4] This law sets out the basic characteristics of pollution-control policy based on a system of pollution permits and pollution charges. It also specifies the responsibilities of the government and the Ministry of Environment for the development of specific regulations to implement the act.[5]

---

dards, effluent/emission standards, supervisory system, enforcement and punishment procedures. For example, environmental impact assessments and land-use regulations (zoning) will play an important role in addressing and managing environmental risks associated with industrial accidents and pollution concerns. A permit and charge approach cannot easily be applied to certain key types of polluters, most notably mobile sources.

4   The exact translation between Estonian and English has not yet been officially determined in Estonia. There is no official translation of Resolution 244. A direct translation is 'Act of Compensation for Pollution Damages' but 'Act on Pollution Charges' better represents the spirit of the law.

5   There are several regulations related to the implementation and structure of this law. See the yearly Estonian environment reports (for example, Estonian Environmental Information Centre, 1994, 1995) for existing and new environmental laws and regulations.

The stated purpose of the act is to restrict disposal of polluting substances into the environment through economic measures and to obtain additional revenue to finance environmental protection. As will be seen, however, these charges are largely designed to enforce compliance with pollution limits specified in enterprise pollution permits. It is also hoped that, by making polluters pay for polluting the environment, there will be some additional incentive for cost-effective pollution control, including some technical changes and better 'housekeeping'.[6]

It is perhaps best to summarize first the basic structure of this system of permits and charges, specified both in this umbrella law as well as in its related regulations, before moving on to further details of the law and regulations. While the details vary somewhat depending on the nature of the pollutant (air, water, solid waste, sea), the basic characteristics of pollution policy in Estonia can be summarized as follows:

- County Environmental Departments (CEDs) are generally responsible for implementing and enforcing the law, including the collection of charges;
- enterprises are responsible for requesting from their CED a permit for specific amounts of different pollutants;
- based on this requested amount, a pollution permit may be granted to the enterprise for a fixed period of time (1–5 years, usually);
- in the permit, threshold levels of pollutants (tons per year or quarter) are defined and concentration limits (grams per second, milligrams per litre) are defined for some situations as well;
- the enterprise pays a constant unit pollution charge for each unit of pollution up to the threshold amount in tons per period (the base charge), and the enterprise pays a higher constant unit charge for those units of pollution above the threshold level (the penalty rate);
- an enterprise does not violate any law by polluting over the threshold level, it just must pay a higher rate;
- if pollution occurs without a permit, such as through some accidental release, the penalty rate is assessed on all units of pollution; and
- the enterprise is responsible for reporting its pollution amount quarterly and paying the required pollution charges to the CED.[7]

6 The purpose of the charges is not to bankrupt existing enterprises producing with outdated and depreciated technologies. There is some hope that by paying pollution charges, polluters will become more aware of environmental laws. It is hoped that the acceptance of polluters to pay such charges is enhanced by the recycling of revenues into environmentally-related actions.

7 There are sanctions for late payment based on 0.2 per cent daily interest rate for late charges. After three months, the case can be taken to a court (administrative court) for a decision.

While there are several detailed differences between types of pollutants that will be identified and discussed later in the text, Estonia in general has a two-step constant charge system based directly on pollution. So, for example, let Ma represent actual pollution, Mp represent the limit in the permit, Cb represent the base charge, and $C_p = nC_b$ represent the penalty charge which equals the base charge times a multiple factor n. Using this notation, the polluter pays $Cb*Ma$ in charges if $Ma \leq Mp$, and the polluter pays $Cb*Mp + Cp*(Ma - Mp)$ in charges if $Ma > Mp$. If the polluter has no permit for polluting a certain substance, either because the polluter did not acquire one when it should or because there was some accidental release, the penalty rate is paid on all units of pollution. The larger the factor 'n', the more costly 'bad behaviour' becomes. The factor n varies from 5 up to 500 depending upon the class of hazardousness of air-polluting substances or stored solid wastes, and n varies from 5 to 10 in the case of water effluents. It is hoped that this two-level charge system can create incentives for enterprises to pollute within limits and to have the necessary permits.

These pollution charges are the main revenue source for the off-budget Estonian Environmental Fund. Within the fund, the revenues are split 50/50 between the central fund and the respective county funds. For reference, in 1994, about EEK 20.6 million were collected in pollution charges, which included both base and penalty charges. In per-capita terms, this is about EEK 14 per person per year (or a little over $1). Pollution charges are paid quarterly and are indexed to the previous year's consumer price index. For example, beginning in 1996, the base charge rates increased 28.1 per cent given changes in the consumer price index between November 1994 and 1995.

Besides the basic charge structure, the Act on Pollution Charges contains two other clauses that are relevant here. First, the act contains a *strict liability* clause, where enterprises remain liable for paying compensation equal to pollution damages to another party due to the enterprise's pollution. Having a permit, and operating within the threshold levels does not eliminate this liability (as a *negligence* approach might imply). Currently, this clause seems to receive little attention, perhaps due to the vagueness of the definition of damages, burden of proof and the generally underdeveloped legal system. It is likely, however, that this clause will gain importance in the future as the legal system develops.

Second, a clause in the act states that an enterprise can receive a pollution charge allowance (or waiver), a decision made by the Minister of Environment, to help pay for investments that would be expected to reduce pollution by at least 25 per cent from the previous year's pollution level. If the pollution charge allowance in the first year is not sufficient to cover investment costs, pollution charges can be reduced for up to three years up to the point where the nominal investment value equals the nominal valuation of the pollution charge allowance.

A pollution charge allowance is directly consistent with the polluter pays principle, where polluters pay for abatement costs but not additional charges, and could be one way to combine a 'stick' and 'carrot' approach to pollution-control policy. Of course, whether a pollution charge allowance alone actually creates any investment depends on whether the investment costs are less than this reduction in pollution charges plus additional reductions in charges granted in the pollution charge allowance.

There are many potential problems with implementing a pollution charge allowance. For example, it does not necessarily create any incentive for an enterprise to find the least-cost way of reducing pollution. It is also likely that an allowance will tend to be granted to finance end-of-pipe clean-up technology, which is easy to identify as an environmental investment, rather than direct win–win investments in enterprise production and management efficiency, which reduces pollution as a byproduct of reducing costs and increasing profits. The Ministry of Environment is aware of this potential problem. Time will tell if this problem can be avoided.

It is important to mention the relationship between the general tax system, pollution charges and incentives for pollution control.[8] Given Estonia's Income Tax Act (Resolution No. 237 of 21 December 1993; in force 1 January 1994), enterprises pay a 26 per cent tax on 'taxable income', where taxable income is defined in relation to gross income and deductible operating expenses (called entrepreneurship expenses). Under Article 16 of the Income Tax Act, pollution charges are not a deductible operating expense for income tax purposes. One main implication of pollution charges being paid out of after-tax profits is that the real unit cost of pollution charges for an enterprise is about 35 per cent higher than

---

8   Regarding other taxes, an excise tax on engine fuel was introduced on 1 July 1993. On 15 November 1995, the motor-fuel excise tax rates rose by 200 per cent, which accounts for about 25 per cent of retail price. It has not been possible to differentiate excise tax rates by the hazardousness of the fuel (content of lead and volatile organic substances in the fuel). There is also an excise tax on motor vehicles (introduced on 1 April 1995). Different rates have been established for imported and locally-produced cars depending on their age and the car's engine capacity (cylinder volume). There is no environmental logic to this tax. For example, for the tax based on age, new vehicles pay the same tax as 10-year old vehicles (the excise tax is 1,000 Estonian kroons (EEK)).

the pollution charge rates specified in the regulations.[9] If pollution within threshold levels in permits is considered to be a normal part of the production process, this part of the pollution charge should be listed in deductible operating expenses. Charges on pollution above the threshold level in the permit, or without permits such as accidental pollution, could continue to be paid from after-tax profits. In reality, this concession on pollution charges being paid from after-tax profits was necessary to get the pollution charges law approved by the parliament. However, Estonia missed a good chance to integrate environmental concerns more directly into the broader tax code.[10]

The income tax law also allows a 40 per cent annual depreciation allowance on assets (Clauses 17(2) and 17(4)). Thus, this 40 per cent depreciation allowance combined with a 26 per cent income tax rate reduces total investment costs by about 10 per cent through its impact on taxable income if the enterprise has taxable income. For many enterprises, if there is some level of taxable income, the value of this depreciation allowance would be much larger than the value of any pollution charge allowance, and much easier to acquire as well.

A second general tax law of interest is the Act on Value-Added Tax (passed in August 1993), with several amendments since, which specifies a basic 18 per cent tax on the selling price of goods and services. Under Article 5, Clause 1, goods and services sold for 'treating dangerous waste' or packaging that is recycled are exempt from this 18 per cent tax. This seems to be a very important clause for incentives to purchase 'goods and services' for 'treating dangerous waste.' Anything that reduces the cost of pollution control by 18 per cent must be considered very important. Unfortunately, the law does not define exactly the term 'treating dangerous waste.' Although pollution is by definition a waste and potentially dangerous to human health and the environment, otherwise there would be no need for any government policy, the current definition is not very clear and could be read in a very exclusive way. Thus, currently most goods and services purchased for pollution control include an extra 18 per

---

9   The 35 per cent number comes from the fact that a profit-maximizing enterprise will make decisions (that is, the first-order conditions for profit maximization) based on the after-tax pollution charge, which is $1/(1 - 0.26) = 1.351$ higher when paid out of after-tax profits.

10   It seems likely that, given the current economic circumstances in Estonia, some enterprises will argue that they cannot pay charges due to operating losses (no profits). The Pollution Charge Act does not specify what to do when enterprises do not have any 'after-tax income' as defined in the income tax law. As a related point, Article 21, Clause 1 in the tax laws states that losses occurred in any year (a situation of negative taxable income) can be carried forward for up to five succeeding periods of taxation.

cent VAT charge. Trying to define clearly pollution control activities that could be exempt from this VAT charge would help to create an economic incentive for pollution control activities, by reducing the costs of these activities.[11]

2.2 Basic Charge Rates[12]

As required in the Act on Pollution Charges, the Government Regulation on Pollution Charge Rates No. 142, passed by the Government of Estonia on 29 March 1995, specifies the basic charge rates and penalty multiples for several types of air and water pollutants and solid waste. Tables 3.1, 3.2 and 3.3 report these pollution charge rates for 1995.

The base rate charge is quite low and for some substances it is probably unreasonably small. Enterprises do not find it difficult to pay charges within limits. At the same time, the penalty rate for some pollutants or for emissions without permits could be very costly. As will be discussed later, Estonia remains flexible in its implementation of these charges in specific circumstances.

Coefficients have been established to adjust pollution charge rates to take into account the local level of pollution, the sensitivity of the local environment and the potentially higher danger to the environment and human beings. These locational coefficients vary between 0.3 and 3.0. As reported earlier in this chapter, the Ministry of Environment together with the Ministry of Finance must adjust the base rates according to the yearly change in the consumer price index from the previous year.

For all types of air pollutants, the penalty charge for pollution amounts over the threshold level defined in a permit are a multiple of the base charge depending on the pollutant's hazard class (four hazard classes). For reference, these multiples are: 5 for the fourth hazardous class; 50 for the third and second hazardous classes; and 500 for the first hazardous class. In this regulation, it is also specified that all emissions occurring without a permit are assessed the penalty rate.

For most types of water pollutants, the penalty charge over the threshold level is multiplied by a factor of five. If an enterprise is without a pollution permit then the total amount is calculated at the corresponding

11 The Government of Estonia (Regulation No. 66 of 18 February 1994) gave a VAT exemption to goods imported to Estonia within the framework of projects financed by loans taken or guaranteed by the Estonian state, as well as environmental equipment and technology imported by the Ministry of the Environment.
12 User charges have existed in Estonia for many years. They are called a 'Payment for Environmental Services'. For example, sewage treatment costs are added to residential and commercial water tariffs.

Table 3.1 Base rates for major pollution charges for air emissions

Pollutant	Permitted	Above permit	Non-permitted
SO_2	20.80	1,040.00	1,040.00
NO_x	47.60	2,380.00	2,380.00
CO	3.00	15.00	15.00
Non-toxic dust	14.90	74.50	74.50
Soot	29.80	1,490.00	1,490.00
Oil-shale ash	20.80	1,040.00	1,040.00

Notes: Set January 1995.
EEK/ton (EEK 8 = DM 1 or about EEK 11 = US$1)

Source: Regulation No. 142 of the Government of Estonia of 29 March 1995 (from Appendix 1).

charge rate multiplied by a factor of 10. Shipping activities which emit oil products into the sea are charged the base rate for oil products times a factor of 50. Heavy metals and other toxic substances which are over the permitted amounts or are without a pollution permit are assessed the maximum charge base rate associated with phenols of EEK 5,920 per ton times a factor of 1,000.[13]

Table 3.2 Base rates for pollution charges for the discharge of pollutants into water bodies, ground water and soil (set April 1995) (EEK/ton)

Pollutant	Permitted	Above permit	Non-permitted
BOD_7	850.00	4,250.00	8,500.00
Suspended solids	430.00	2,150.00	4,300.00
Phosphorous	1,280.00	6,400.00	12,800.00
Nitrogen	710.00	3,550.00	7,100.00
Oil products, shale oil	1,700.00	8,500.00	17,000.00
Sulphate	10.00	50.00	100.00
Phenols (mono-)	5,950.00	29,750.00	59,500.00

Source: Regulation No. 142 of the Government of Estonia of 29 March 1995 (from Appendix 2).

13 Charges are also assessed on the pH level of effluent if it is above or below the pH range of six to nine (EEK 0.14 per 0.1 pH unit/m^3 of effluent).

Table 3.3 Base rates for pollution charges for the disposal of waste into
the environment (depositing and burying; set January 1995)
(EEK/ton)

Classification of waste based on their level of hazardousness	Permitted	Above permit	Non-permitted
Extremly hazardous (I class of hazardousness)	595.14	297,570.00	297,570.00
Hazardous waste (II class of hazardousness)	59.51	5,951.00	5,951.00
Moderate hazardousness (III class of hazardousness)	5.95	297.50	297.50
Minor hazardousness (IV class of hazardousness)	2.13	10.65	10.65
Inert wastes	0.71	3.55	3.55

Source: Regulation No. 142 of the Government of Estonia of 29 March 1995 (from Appendix 3).

For solid waste, if an enterprise does not have a pollution permit or its actual amounts are above its permitted amounts then charges are multiplied by a 'penalty' factor which ranges from 5 to 500 according to its hazardous class.

2.3 General Issues

The basic economic theory of pollution control is well known, and Estonia's mixed system of permits, threshold levels and pollution charges designed primarily to enforce compliance with threshold levels does not neatly fit the basic theory.[14] While there is a large literature comparing when and why economic instruments are preferred to various forms of direct regulation, the practical difference for Estonia between compulsory 'command-and-control' and the 'market based' is perhaps not so great. The reasons have to do with three issues: (1) the number of important polluters; (2) the location of the polluters; and (3) the difference between uniformly-mixed and non-uniformly-mixed pollutants.

14 Tietenberg (1992a, Chapter 12) provides a thorough introduction to the topic.

Most of the point-source pollution in Estonia is caused by a few large enterprises. According to the Estonian Environmental Information Centre in 1994, the ten largest out of 807 point-source air polluters accounted for 92 per cent of solid particulates emissions, 85 per cent of sulphur dioxide (SO_2) emissions and 73 per cent of nitrogen oxide (NO_x) emissions. The two worst polluters of these ten, the electric power plants burning oil shale, contribute 53 per cent of particulates, 70 per cent of SO_2, and 60 per cent of NO_x. Also in 1994, water effluent discharges of the three biggest cities in Estonia – Tallinn, Tartu and Narva – accounted for 50 per cent of BOD_7, 73 per cent of oil products, 37 per cent of phosphorus (Ptot) and 42 per cent of nitrogen (Ntot). The technologies of these polluters are either almost the same, so that their marginal abatement costs are essentially the same, or the polluters emit into different environments without a uniform environmental impact.[15] Because of the differential environmental impact of these large polluters, economic theory would suggest that separate tax rates are needed for each location and/or enterprise. Estonia has attempted to solve this problem by using a combination of permits with threshold levels and some locations weights to adjust basic charge rates.

Given the structure of pollution charges explained in Section 2.1, the enterprise's structure of pollution-abatement costs (perhaps better thought of as its marginal benefits from pollution generation) determines if the two-step charge system actually creates any incentives to be at or below the threshold level in the permit. It is perhaps important to emphasize here several types of information that must be known for such incentives to be created. Perhaps the most important assumptions are that pollution is easy to define and monitor, and marginal abatement costs and marginal damage costs are known by the enterprise and the environmental policy authority.[16]

If enterprises do not have good information available so that they know their marginal abatement costs, then charges at least in the shorter run create little incentive to alter pollution in a least-cost manner. And, since

15 For example, effluent in the north of the country from the Tallinn Municipal Sewage Treatment Plant into Muuga Bay of the Baltic Sea has a very different environmental impact from effluent in the southeast of the country from the city of Tartu into the Ema river.
16 Pezzey (1988) provides a good summary of underlying assumptions of this type of model. For policy based on pollution standards, charges, or some combination of both, much of the same type of information is needed for the policy to be effective. If just a standards approach is followed, it is still necessary to determine the standard, monitor emissions, and enterprises need to know abatement costs. Marginal damages must also be known to follow a tax approach based on equating marginal abatement costs and marginal damages.

the efficient level of pollution depends directly on the firm's marginal benefits from pollution (marginal cost of pollution abatement), a 'correct' Pigouvian tax does not necessarily imply a large – or any – reduction in emissions for an existing marginal abatement cost function.[17]

It is noted here that a *polluter pays* approach to environmental policy, recommended by the OECD and part of European Union environmental policy, is directly consistent with a standards approach but is not clearly related to a charging system. Following Pezzey (1988), the basic polluter pays approach implies that firms are responsible for paying their *costs of pollution abatement,* which happens with a pollution standard approach. With a charge, polluters pay these abatement costs and the pollution charge. With an extended version of the polluter pays principle, the firm could also be charged for direct pollution damages. This payment for damages implies, essentially, a *strict-liability rule* for environmental policy. However, neither version of polluter pays is directly related to a Pigouvian charge system, where all units of pollution are assessed the charge. In principle, as Pezzey (1988) suggests, a simple two-step pollution charge as used in Estonia can provide the right incentives for pollution control while reducing the tax burden on enterprises.

From a public finance point of view, there is not necessarily anything wrong with a Pigouvian tax that collects more revenues than social damages, since this tax creates the right social marginal incentives in the economy. In fact, environmental taxes could be more clearly integrated with non-environmental taxes to achieve a less distortionary tax system. Perhaps an eventual outcome of such integration would be a system where environmental taxes could provide incentives to reduce pollution and provide revenues so that other more distortionary taxes could be reduced. This point follows directly from the OECD report on taxation and environment, which concludes that 'there are substantial gains to be made from the integration of taxation and environmental policies' (OECD, 1994, p. 12).

Estonia hopes that the current system creates incentives for innovation, through investments in better management and technology, to reduce the costs of pollution regulation on the enterprise.[18] Whether any

17 If the charges are large for such enterprises, then they may be inclined to abate in a relatively costly manner with unnecessarily large economic and perhaps social impacts. If the base charges are relatively low and the threshold levels relatively lenient to begin with, as is the case in Estonia, it is likely that actually paying charges starts enterprises thinking about their pollution levels and starts them exploring cost-efficient pollution-control strategies.
18 See Tietenberg (1992a, Chapter 14) and Pezzey (1988) for a more complete discussion of incentives for pollution control under pollution charges or standards for this type of simple model.

investment actually takes place depends on whether the investment costs are less than the benefits to the enterprise. Just because a tax creates more incentives, it may not create enough incentives for enterprises to pay for the investment, particularly given the often thin credit markets available to enterprises in transition economies. In effect, then, raising charges does not create an automatic incentive to invest and could, in fact, reduce an enterprise's ability to invest if it was also paying higher charges.

While Section 2 provides a brief overview of Estonia's mixed system of permits and charges, there are some important differences between the structure of air and water. To look at these differences and to evaluate recent implementation experience more closely, Section 3 focuses directly on air emissions and Section 4 focuses on water effluents.

3 AIR EMISSIONS

3.1 Permits, Ambient Standards and Threshold Levels

Air pollution permits are generally issued by County Environmental Departments. A pollution permit is required if the annual emission of a pollutant is above some minimal amount stated in Estonian regulations. In the case where an enterprise applying for a permit has an air pollution source which is greater than 100 metres, the permit will be issued by the Ministry of Environment, which will have received a written opinion from the CED.

There is currently a link between ambient air quality goals and threshold levels in permits. The pollution charges for air pollutants and the guidance on threshold levels are based on a list of Estonian ambient air quality standards (the maximum permissible ground-level concentrations of the air pollutants). There are two types of ambient standards: a 24-hour average maximum permissible concentration (LPK_k is the acronym in Estonian) and a one-time (30-minute) maximum permissible concentration (LPK_m). Table 3.4 reports ambient standards for several key air pollutants.

These standards exist for about 139 different pollutants organized into four hazard classes. The permissible pollution concentrations are based on the norms of the former Soviet Union, but they are changing step by step grounded on the experience of other foreign states and on recommendations of the EU and the WHO. For example, as discussed in Ensmann et al. (1995), the ambient standard for NO_2 nitrogen dioxide (NO_2) was too strict (LPK_m = 0.08 mg/m^3 and LPK_k = 0.04 mg/m^3). Based on further consideration and recommendations from the WHO, a

Table 3.4.   Ambient air quality standards (micrograms per cubic metre)

Pollutant	LPK_m	LPK_k	WHO	EU	US
SO_2	500	50	125	100–150	365
CO	5,000	3,000	10	–	10
Non-toxic dust	500	150	100–150	80–130	50
NO_2	400	150	150	50–200	100

Source: The standards reported for the EU, the WHO and the US are taken from UNEP/WHO (1992). These standards, while based on some form of a 24-hour average, are not exactly comparable due in part to definition of pollutant and sampling method. The exact details of differences are explained in the UNEP/WHO report.

ministerial regulation of 7 September 1995 established new ambient standards for NO_2 reported in Table 3.4 (LPK_m = 0.4 mg/m^3 and LPK_k = 0.15 mg/m^3). There continues to be room to reevaluate Estonian ambient air quality standards. For example, the standard for SO_2 may be too strict while that for CO may be too lenient.

LPK_m is used to define the threshold level in the pollution permit. For reference, the process to receive a threshold level in an air permit can be divided into three steps. First, the enterprise requests a threshold level of pollution in its permit. Presumably, this request is based on expected production plans and input use for the year. Any rational enterprise manager would request some amount extra for insurance in case production is larger than planned. Second, based on the emission rates implied by this level of pollution, a dispersion model of ambient air quality is supposed to be used to determine whether this emission rate would lead to a violation of the LPK_m standard at any time during the year. In principle, various conditions used to parameterize the model, such as temperature, wind speed and emissions from other polluters, are based on worst-case scenarios. And third, as long as LPK_m is not estimated to be violated for the worst-case scenario, the threshold level requested by the enterprise is approved. Approximately five dispersion models have been 'approved' by the Ministry of Environment for conducting these air quality modelling exercises. The quality or validity of these models is not clear.[19]

19   It is emphasized here that permits specify emission standards. Command-and-control technology standards do not exist at this time. This system is somewhat inconsistent with the Helsinki Convention, which has been ratified in Estonia. The Helsinki Convention requires that, when issuing permits for new pollution sources, a best available technology approach should be followed.

This worst-case scenario approach to giving air permits may need some reevaluation. For example, it is possible that the maximum would be expected to be violated only a very small amount of time during a year, but in general a large number of days would be very good. However, this permit would be rejected. On the other hand, another firm could request a permit where the maximum is not expected to be exceeded, but a large number of days would be close to the maximum, and this permit would be approved. In terms of human health, it is not automatically clear what is a better policy. In should be noted that in some areas, such as the European Union, ambient air standards explicitly incorporate this probabalistic nature of pollution directly into the standard.

3.2 Additional Air Emission Charges

Also in Governmental Regulation No. 142, a formula based on ambient standards is specified to determine base charge rates for all pollutants not specified directly in Table 3.1. For reference, this formula is:

$$T = \frac{k}{\sqrt{LPK\_m * LPK\_k}}$$

where T = rate of pollution charge in Kroons per ton; LPK_m = one-time maximum permissible concentration of pollutant in ambient air, milligrams per cubic metre; LPK_k = 24-hour average maximum permissible concentration of pollutant in ambient air, milligrams per cubic metre; k = EEK 3.3, $\times$ mg/t $\times$ m³. As will be shown in Section 3.3, such charge rates from (3.1) are relatively not important within the total charge system.

3.3 Implementation Experience

The data on Estonian pollution generation and pollution taxes parallel the concentrated geographic distribution of heavy industrial production in the country. In sum, a few large, primarily state-owned enterprises in the northeast of Estonia account for most of the officially recorded pollution and collected pollution charges in the country (Ensmann et al., 1995).

For 1994, the 172 polluters in the northeast and Tallinn, out of 807 nationally, accounted for 249,900 tons of all types of air pollutants out of the 354,000 total estimated for the country. Of this total level, the regional/national amounts of key pollutants include: 92,700 tons of solids (particulates) out of 161,500; 124,700 tons of the SO_2 out of 141,100; and 12,000 tons of the NO_x out of 14,600. In other words, the northeast and

Tallinn account for about 70 per cent of total air emissions, 57 per cent of particulates, 88 per cent of SO_2 and 82 per cent of NO_x (Estonian Environmental Information Centre, 1994, p. 67).

While the majority of air emissions occur in the northeast and Tallinn regions, it is also important to note that, of the 172 air polluters in these regions, the top ten point sources account for the majority of air emissions and paid pollution charges. For example, nine of the ten largest air polluters in the country were located in this region (and the other largest, Kunda Nordic Cement factory, was close by as well).

Figure 3.1 provides information on the concentration of pollution charges across enterprises, while Figure 3.2 reports the concentration of charges across pollutants. As one can see from Figure 3.1, the two state-owned electric power plants (the Estonian Thermal Power Plant and the Baltic Thermal Power Plant) accounted for more than 70 per cent of all charges paid, with the Tallinn Municipal Heating Enterprise, the Kohtla-Järve Thermal Power Plant, the Iru Thermal Power Plant and the Kiviter Chemical Company accounting for another 10 per cent of charges. All of these polluters are under some form of public ownership. The remaining 166 enterprises in the regions accounted for 14 per cent of charges.

From Figure 3.2, SO_2 and ash (from burning oil shale for electric power plants) accounted for more than 82 per cent of all charges in the area and NO_x accounted for about 12.5 per cent of charges, even though charges were actually collected on about 70 different air pollutants. Thus, for most practical purposes, the charge rates specified in the governmental regulation are the most important pollution charges. While equation (3.1)

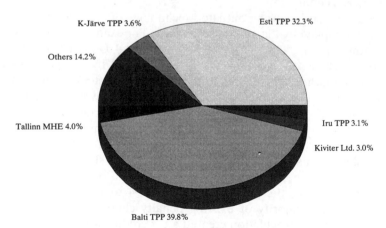

Figure 3.1   Air pollution charges by polluters as percentage of total charges in Tallinn, Narva and Ida-Viru in 1994

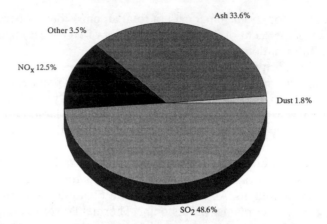

*Figure 3.2   Air pollution charges by pollutants as percentage of total
charges in Tallinn, Narva and Ida-Viru in 1994*

provides the rates for a large number of other charges, there is little
total importance of such charges. It is very possible that the transaction
costs associated with actually implementing charges on these 'other' 70
pollutants – monitoring, reporting, filling out forms, entering data into
computers, collecting the charges – are much higher than any environmental benefit from imposing the charge. At the same time, it is also the
case that, given the human capacity constraints within the CEDs, their
attention would be better spent for now on a limited quantity of larger
polluters in each county.

In practice, enterprises using outdated technology often exceed emission standards implied by the ambient standards. For example, according
to the information of Lääne-Virumaa Environmental Department, the
emission of oil-shale ash exceeded the threshold level by 660 per cent
in 1994 in the boilerhouse belonging to RAS (state joint-stock company)
Kunda Tehased in Kunda town, which also supplies town inhabitants
with hot water and heating. For such an excess RAS Kunda Tehased
should have had to pay a pollution charge worth 2.4 million kroons.
As a result, in the beginning of 1995, the Governmental Regulation
specified pollution charge rates were modified. For oil-shale ash, a ten
times smaller pollution charge penalty rate (5 instead of 50) was temporarily established for the period 1 January 1994 to 1 January 1997.
This eliminated the majority of existing penalty charges during 1994.
In effect, this change in regulation created a tax credit for the company,
in return for which the company promised to transfer to gas heating
by the end of 1996.

As another example, actual emissions of clinker ash in the Kunda Nordic Cement plant exceeded the permit threshold level by 900 per cent, while cement ash emissions exceeded the threshold level by 360 per cent. In 1994, the pollution charge payment for clinker and cement ashes was EEK 25 million, but the company did not pay because at the beginning of 1994 the Minister of Environment and the company concluded an agreement to allow a pollution charge allowance of EEK 27 million.

4　WATER EFFLUENTS

Since 1991, the system of water-pollution permits and charges has evolved through three stages. Each stage is discussed below and the main implications of these changes are discussed.

4.1　Permit Process and General Charge System During 1991–1993

According to the Water Act a user of water must have a permit for special water use. This permit is for a fixed term, and it is issued by a County Environmental Department. The permit is based on the Procedures for Issuing Water Permits (Ministerial Regulation No. 2, passed by the Minister of Environment on 28 February 1994). This regulation gives several responsibilities for implementation and enforcement to the CEDs, including issuing permits, determining threshold levels and time period of permit, and some monitoring of pollution levels.[20]

A permit for special use of water consists of two parts: water use/consumption and discharge (pollution permit). The discharge permit specifies: (1) maximum permissible concentration (mg per litre) of specific pollutants; and (2) maximum total permissible discharge of a pollutant (tons per quarter, and kg per 24 hours, if necessary). The details must be written into the permit as to how, when and where enterprise self-monitoring needs to be conducted.

For water, there were not and are not ambient surface water-quality standards. During this period, CED staff largely determined threshold effluent levels, both concentration levels and tons per year. The permits, as a result, were region and enterprise specific. The process was perhaps best described as a negotiation. The enterprise was supposed to conduct a study or have a study done on its production process with its existing technology to determine the amount of pollutants generated as a result

---

20　The Water Act contains clauses as to when a CED may refuse to issue a permit or when a permit can be invalidated. However, the governmental regulation needed to determine invalidation procedures has not yet been approved.

of its production expectations for the year. The enterprise then requested in its permit a threshold level that would satisfy this expected amount, and, if it wanted to be safe, it asked for some additional amount. A negotiation process then ensued between the CED and the enterprise on the threshold amount, perhaps taking into account the enterprise's feasible technology based on what was already in existence. Since CEDs did not have a clear methodology for determining permitted levels, concern existed over how stringent permits would stand up in court.

One byproduct of this lack of threshold guidance is that the CEDs were free to negotiate a compliance schedule with the enterprise, where the permit could be written in a way to give the enterprise time to adjust to tighter threshold levels in the future. This compliance schedule could also be built into the schedule of permit applications and renewal over time.

At the beginning of 1991 until the end of 1993 the simple double-staged system explained in Section 2.1 was valid. At the time, there was serious concern about decisions made surrounding permits in terms of effluent limits and other details in the permitting process.

4.2   New Standards and Revised Charge System During 1994–1995 (First Quarter)

A major change in Estonian policy occurred in 1994 when effluent concentration standards were established following HELCOM recommendations.[21] The effluent standards were first informally adopted and then were established officially by the government in December 1994. The effluent standards are specified for main water pollutants depending on the average daily quantity (flow rate) of wastewater. The effluent standards are reported in Table 3.5.

In some cases, the HELCOM effluent standards are minimum requirements. When originally adopted, if the standards did not achieve a sufficient ambient quality of a water body (somewhat loosely defined at the time), the CED had the right to strengthen standards. However, according to a governmental regulation of 15 December 1994, these standards can be tightened only by the government on application by the Minister of Environment. But a newer governmental regulation of 25 April 1995 gave some rights to CEDs: (1) to allow the establishment of tighter requirements in wastewater treatment in Harjumaa and Järvamaa, whose

21   HELCOM (Helsinki Commission) was established by the Baltic Sea countries to reduce pollution loads in the Baltic Sea. Related to HELCOM, Estonia signed the 'International Water Protection Cooperation Agreement' with Finland on 9 April 1992, in which Estonia agreed to several conditions.

Table 3.5   Effluent concentration standards (average mg/litre)

Pollutant	Over 2,000* m³/day	201–2,000* m³/day
Biochemical oxygen demand (BOD₇)	15	25
Suspended solids	15	25
Total phosphorus (Ptot)	1.5	2
Phenols	0.1	not defined
Oil products and shale oil	5.0/1.0**	5.0/1.0**
Total nitrogen (Ntot)	10.0***	12.0***

Notes
* By the statistical report Water Use (*Veekasutus*), the sum of average daily wastewater discharge (flow rate) rounded to whole number of water users located within the boundaries of a city or a settlement or the average daily wastewater discharge (flow rate) rounded to whole number of a water user located outside a city or a settlement. For small dischargers, less than 201 m³/day, the BOD₇ suspended solids, and oil products standards are the same as 201–2,000, while standards are not defined for Ptot, Ntot and Phenols.
** For wastewater that has been treated biologically.
*** For modernized and new wastewater treatment plants by 2010.

Source: Regulation No. 464 of the Government of Estonia of 15 December 1994 and Regulation No. 201 of the Government of Estonia of 25 April 1995.

wastewater flows into the Gulf of Muuga or the Pirita river, and on all objects in Valgamaa, whose wastewater flows into Pühajärv, Neitsijärv, the Lake of Köstre, the Elva river and the Pedeli river; and (2) to leave in force more strict effluent standards in permits for those sewage treatment plants operating before 1 January 1995, until new renovations are needed.

There is a some difference between new and repaired sources and existing sources. New sources and repairs must meet the effluent standards immediately, with the exception of total nitrogen which must be met by 2010, while existing sources must meet these standards by the year 1997. It is perhaps interesting to note that the effluent standards were adopted in Estonia with little or no analysis of the implications of the standards. Such regulatory impact analysis could have considered the costs of achieving the standards, the number of sources that will not comply with the standards by 1997, and the Estonian environmental benefits of adopting such standards.

Based on the adopted effluent standards, Estonia then adopted a three-step charge system that alters the basic charge system described in Section 2.1 in a simple way. A zero charge existed for pollutant amounts associated with under-HELCOM standards (determined by multiplying discharge concentration and volume). Between the effluent standards and the limit determined in the permit, the base charge was paid. The penalty charge was paid for amounts above the limit and on all units if an enterprise did not have a permit.

At the time it was hoped that this policy of a zero charge within the effluent standard would provide an additional incentive to meet the effluent standards. In reality, the zero charge provided no marginal incentive for enterprises to reduce effluent levels if they were already emitting above the standard. The reason is that, for example, given a marginal abatement cost function that intersected the base-rate charge at an amount above the effluent standard, reducing per unit charges on some units below this amount has no marginal effect. In effect, this was a free pollution charge waiver without any required investment or effluent changes. Given the relatively small amount of this waiver, however, there was little incentive for technical change as well.

By adopting this policy, Estonia soon faced some problems. For example, the relationship between pollution quantity and pollution charges was effectively broken, and very large amounts of pollutants were discharged without any charges at all. In Tallinn, for example, which is one of the largest water pollution sources in the country, Table 3.6 shows that between 91 and 99 per cent of the total load for major pollutants was discharged into the Baltic Sea and the Pirita river without charges.

As an example using suspended solids, Table 3.7 shows the lack of correlation between discharged amount of pollutants and total charges for six large point sources in Tallinn. For example, Tallinn Waterworks was the largest polluter with more than 271 tons of BOD_7, but paid no charges, while much smaller polluters such as the housing district or the Agro company paid a few thousand Kroons in the first quarter. The Maardu dwelling region emitted less than 10 tons but paid about EEK 7,000 in the quarter. It should be emphasized that both Tallinn and Maardu are emitting into the Baltic Sea directly or almost directly. As a result, there was no environmental damage difference to justify this large payment difference. Needless to say, this somewhat undesirable property of the charge system was eventually changed.[22]

22   Tables 3.6 and 3.7 also show the concentrated structure of water polluters in Tallinn. Table 3.7 shows that Tallinn waterworks accounted for more than 250 tons of suspended solids per quarter, while Table 3.6 reports a total suspended solid discharge of 317 for Tallinn as a whole.

Table 3.6 *Distribution of main pollutants by the charge classes in Tallinn (example data of the First Quarter 1995)*

	Pollution load (ton/quarter)				
Pollutant	Total	Free of charge	Normal rate of charge	Five-fold rate of charge	Ten-fold rate of charge
Suspended solids	317.755	305.856	4.079	7.807	0.01
BOD_7	176.274	163.809	6.068	6.311	0.08
Ntot	311.574	284.171	25.825	1.573	0.00
Ptot	22.963	22.689	0.105	0.168	0.00

Source: Tallinn Environmental Board (1996).

Table 3.7 *Emissions of suspended solids and pollution charges in Tallinn (first quarter, 1995).*

Enterprise	Emissions (tonns)	Charge (base rate) (EEK)	Charge (penalty rate) (EEK)
ME Tallinn Waterworks wastewater purification station	271.400	0	0
Eesti Fosforit Ltd. rainwater outlet	27.920	0	0
Maardu Dwelling Maintenance Enterprise outlet from industrial region	3.601	0	7184
The Patarei Prison	3.068	1180	0
Dwelling region Suur-Patarei	2.240	136	3273
Agro Plantation of ECC	2.138	4	4338

4.3   Revised Charge System as of 1995 (Second Quarter)

The water charge system changed for a third time in 1995, where the three-step system discussed in Section 4.2 was revised. Recall the notation from Section 2.1 where Ma represents actual pollution, Mp represents the limit in the permit, Cb represents the base charge, and $Cp = n*Cb$ represents the penalty charge which equals the base charge times a multiple factor n. The new water system also includes a discharge concentration (mg/M^3) parameter X, where Xa represents actual concentration and Xh represents the concentration standard if it exists in Table 3.5 or the concentration limit specified in the permit for other pollutants.

Based on this notation, the new system now includes three possibilities. First, if $Xa \leq Xh$ and $Ma \leq Mp$ and all other conditions in the permit are satisfied (monitoring, reporting, and so on), then the polluter pays $0.5*Cb*Ma$. Second, if $Ma \leq Mp$ but $Xa > Xh$ and/or other conditions in the permit are violated, the polluter pays $Cb*Ma$. And third, if $Ma > Mp$, then the polluter pays $Cb*Mp + Cp*(Ma - Mp)$. Thus, as opposed to the previous system described in Section 4.2, all units of pollution are now assessed some charge in the new system even if the concentration standards are satisfied. The second and third cases revert back to the original system described in Section 4.1.

As reported in Tables 3.8 and 3.9, and as compared to Tables 3.6 and 3.7, the main impact of this change in charge structure has been to increase charge revenues and recreate a clearer correlation between

Table 3.8   *Distribution load of main pollutants by charge classes in Tallinn (example of the Second Quarter 1995)*

	Pollution load (ton/quarter)				
Pollutant	Total	0.5-fold rate of charge	Normal rate of charge	5-fold rate of charge	10-fold rate of charge
Suspended solids	312.615	297.262	6.312	8.967	0.074
BOD7	168.031	156.201	4.810	6.950	0.070
Ntot	301.445	293.072	6.639	1.728	0.006
Ptot	13.488	12.798	0.526	0.163	0.001

Source: Tallinn Environmental Board

Table 3.9   Emissions of suspended solids and pollution charges in Tallinn (second quarter, 1995).

Enterprise	Emissions (tonns)	Charge (0.5 rate) (EEK)	Charge (base rate) (EEK)	Charge (penalty rate) (EEK)
ME Tallinn Waterworks wastewater purification station	281.400	72601	0	0
Eesti Fosforit Ltd. rainwater outlet	12.502	0	8064	0
Eesti Fosforit Ltd. biological treatment plant	3.867	0	2296	993
The emergency outlet of ME Tallinn Waterworks	3.713	0	266	10643
Maardu Dwelling Maintenance Enterprise outlet from industrial region	3.110	0	143	9317
Iru TPP	2.174	701	0	0

pollution quantity and pollution payments. Since Tables 3.7 and 3.9 report the six largest polluters in Tallinn in each quarter, the polluters change between tables. As Table 3.8 shows, all amounts now receive some charge. From Table 3.9, the Tallinn sewerage treatment plant now pays the highest charge in relation to other sources, in contrast to Table 3.7, which is more in line with its pollution level.

The new water-charge system now creates some non-marginal changes in the structure of water pollution charges. Essentially, the polluter can choose between two charge systems based on concentration level, pollution quantity and satisfying other conditions in the permit. As a result, it is now possible that the water-charge system does create some incentive for enterprises to emit within Xh and Mp to pay the reduced base rate. While it is too early to tell if such changes will actually be created, there are some indications that some enterprises have undertaken some 'good housekeeping' actions that have substantially reduced effluent concentration levels. For example, comparing effluent concentration levels for suspended solids for the top 20 polluters between the first and second quarter of 1995, which corresponds to the end of the old system and the beginning of the new system, effluent concentration levels essentially remained constant and/or within concentration standards for eleven enterprises, fell substantially for six enterprises, and increased somewhat for four enterprises (Tallinn Environmental Board, Personal Communication, 1996). One clear example involves a company where effluent concentration of suspended solids fell from more than 300 mg/litre to 4 mg/litre between the first and second quarter of 1995 (Tallinn Environmental Board, Personal Communication, 1996).

5 SUMMARY AND RECOMMENDATIONS FOR THE FUTURE

It is well recognized in Estonia that implementation of pollution-control policy is not as effective as it could be. This is not unexpected. The laws and regulations are relatively new, and time and experience continues to be needed to work out any problems, and to learn from mistakes and the analysis of the impacts of the current system. Several examples, some of which have been discussed in this chapter regarding revised NO_x standards and the charge structure of water effluents, show that such learning and revision can and does take place. For the longer term, it is clear that Estonia seeks integration with the European Union, which will imply closer eventual harmonization of Estonian environmental policy with European Union policy.

Based on this analysis and the authors' first-hand knowledge, there are some main issues on which Estonia can work now to improve upon its basic policy environment. These issues are discussed briefly, below.

First, Estonia should continue to revise its ambient air standards and it should develop some form of ambient surface-water quality goals (for example, standards, goals, use categories). The pollution charge system works in close cooperation with administrative instruments, primarily with pollution permits. Since there can be very large differences between the base rates (for pollution below the threshold level) and the penalty rates (for pollution above the threshold level), this threshold level is key to the pollution permit and charge system. In principle, the threshold level is based on ambient standards for air and effluent standards for water. Ambient air standards for main pollutants need to be revised, as was recently done with NO_x. Some form of ambient surface-water quality goals are needed, whether standards or some other form. If this is done, it will be possible to evaluate if the water effluent standards are actually achieving a result desirable for Estonia. It will also then be possible to know when stricter effluent standards are needed, such as in the case when single or multiple sources are within effluent standards but the environmental impact is still undesirable.

A related issue here is that the standards need to allow for environmental fluctuations by allowing some probabilistic rate of compliance with some limit. Such an approach is common and probably better suits the needs of Estonia now and in the future.

Second, there is a serious question as to whether all potential polluters should be required to have permits. Given the concentration of point-source pollution among a few large polluters, one could question the usefulness of creating and enforcing the large number of pollution permits (about 2,800 permits total for air, water and solid waste) that exist currently in Estonia. Unless data show otherwise, which to our knowledge they do not, there are probably few nets benefits to Estonia of enforcing the permit and charge system for the smaller polluters in terms of costs to the enterprise (monitoring, calculations, studies), costs to the CEDs (administrative, employee work time), and the costs of the small if any reductions in pollution levels as a result of the effort to enforce compliance. Given the limited resources of the CEDs, perhaps greater use should be made of *not* requiring a permit for more enterprises and devoting more energies to the few large polluters.[23]

Third, with a system based directly on pollution quantities, monitoring or measuring the actual pollution amounts remains the key to this entire system of pollution permits and charges. And, now that a relatively large amount of experience has been gained with implementing the system of permits and charges in Estonia, it is perhaps time to focus more attention on monitoring and enforcement. In the present situation, polluters have permits. What has become more difficult is enforcing the conditions/requirements in the permits. At a minimum, the details of exactly how, when and where monitoring needs to be conducted should be written directly into the permit.

Self-monitoring and self-reporting may also be a tenuous situation in Estonia, as in other countries. The enterprise is put in the situation of reporting when it has gone over its threshold levels knowing that this

---

23 There could be some resistance to such changes in Estonia because it is assumed that the information acquired on small polluters is in some way important. Instead of having no permit, it could be possible to request a very simple permit with self-reported pollution levels without accompanying charges.

action will result in substantially higher payments, while the penalties for not correctly reporting are not that much higher. Since it is virtually impossible to calculate pollutant amounts after the fact, except for some types of solid waste, it can be difficult for the CED to determine whether the enterprise truthfully reported.

Fourth, even though the desire is to create increased incentives to pollution control, it is not clear that the current permit and charge system creates incentives for enterprises to control pollution in a least-cost manner. Several assumptions regarding enterprise behaviour and knowledge may not be valid currently for many polluters in Estonia. For a combination permit and charge system to be effective in creating incentives to reduce pollution in a least-cost manner, enterprises must want to maximize profits or at least minimize overall costs, and enterprises must know their costs of reducing pollution. Thus, a prerequisite is a very good accounting system (both technical and financial) in the enterprise. Given that the vast majority of heavy polluters are stated-owned enterprises, these assumptions may not be valid.[24] In effect, then, pollution charges are more like a lump-sum transfer from the enterprise (and perhaps eventually the national and local budget) to the Estonian environmental fund.

And, fifth, while the basic logic of Estonia's permit and charge system follows from a world of certainty, substantial uncertainty actually exists regarding the generation of pollution, environmental damages and abatement costs. Economic and environmental uncertainty (variability, fluctuations) exist at several levels in the system of permits and charges.

The standard economic approach under such uncertainty would be to equate expected marginal benefits with expected marginal costs of abatement (Tietenberg, 1992b). Again, however, to do such calculations it is necessary to have rather sophisticated management capabilities and the

---

24   For example, RAS Kiviter is a large state-owned chemical plant that relies on oil shale as a key input. As of 1993, the enterprise says it has no records of what products it sold and how much. Such poor records, if true, would be astounding for a private enterprise. If it does not know what it sold, it cannot know what it produced.

The least-cost method of pollution control for an enterprise will vary greatly, not only among industries but also among enterprises in the same industry. This type of selection demands specific information on the possible control techniques and their associated costs. Generally enterprises are able to acquire this information when it is in their interest to do so. While it was not in the interest of state industries previously, it is more likely that knowledge of least-costly pollution abatement will be a positive byproduct of the privatization process in Estonia. It should be emphasized that privatization will not solve all of Estonia's pollution problems, but privatization will help to provide several of the preconditions that are needed for a permitting and charge system to be effective. However, with privatization, firms may have a greater incentive not to report all of their emissions or their costs of reducing emissions.

probability distributions for the random variables must be specified to some degree. It is hard to argue that such management capabilities already exist within the largest polluters. It is also hard to argue that probability distributions can be reasonably evaluated for the future within the rapidly transforming Estonian economy. This issue will gain in importance in the future as the ability to monitor and enforce the current system increases, real charge rates increase, and the market economy sharpens managers' need to minimize costs to remain competitive in Estonia's open-market economy.

REFERENCES

Ensmann, E., L. Gornaja and B.A. Larson (1995), 'The Pollution Policy "Cocktail" in Estonia: Economic Incentives and Problems for Implementation', Environment Discussion Paper Series #5, Harvard Institute for International Development.

Estonia Ministry of Economic Affairs (1994). *Estonian Economic Survey, 1993–1994*, Tallinn.

Estonian Environmental Information Centre (1991, 1993–1995), *Estonian Environmental*, Tallinn: Tallinna Raamatutrukikoda.

OECD, DAC (1992), 'The use of Economic Instruments for Environmental Management in Developing Countries', Proceedings of a Workshop held at OECD Headquarters in Paris, 8 October.

OECD (1994), *Taxation and Environment in European Economies in Transition*, Centre for Co-Operation with the European Economies in Transition, Paris: OECD/GD.

Opschoor, J.B. (1993), 'Economic Instruments in Environmental Policies: Why, When, Where?' Paper prepared for Workshop on the Application of Economic Instruments for Environmental Management, United Nations Economic Commission for Europe and the United Nations Environment Programme, Tallinn, Estonia, December.

Pezzey, John (1988). 'Market Mechanisms of Pollution Control: "Polluter Pays", Economic and Practical Aspects', in R. Kerry Turner (ed.), *Sustainable Environmental Management: Principles and Practice*, Boulder: Westview Press, pp. 190–42.

Statistical Office of Estonia (1993), *Statistical Yearbook 1993*, Tallinn.

Statistical Office of Estonia (1995), *Energy Balance*, Tallinn.

Tallinn Environmental Board (1996), Personal Communication.

Tietenberg, T.H. (1992a), *Environmental and Natural Resource Economics*, 3rd edn, New York: Harper-Collins.

Tietenberg, Tom H. (1992b), 'Economics of Pollution Control: An Overview', in *Environmental and Natural Resource Economics*, 3rd edn, New York: Harper-Collins, pp. 360–89.

Timber Industry (1990), 'The State of Environment in USSR in 1988', Moscow (in Russian).

UNEP/WHO (1992), *Urban Air Pollution in Megacities of the World,* World
    Health Organization and United Nations Environment Programme, Oxford,
    UK: Blackwell.
United States Agency for International Development (1994), 'Pre-Feasibility
    Report', Demonstration Environmental Action Programme Project, RAS
    Kiviter, Washington, DC.

4. Implementing Pollution Permits and Charges in Latvia

Janis Brunenieks, Aija Kozlovska and Bruce A. Larson

1 INTRODUCTION

Latvian environmental policy was reborn, with several legs to try to stand on, following independence from the former Soviet Union in 1990. However, these inherited legs were weak and not working together. Since that time, the legs have been developing, new legs have been growing, and slowly it has begun to stand up, to stretch. The Latvian system of pollution permits and charges is one important leg of overall environmental policy.

This chapter begins to analyse the theory and implementation experience behind the Latvian approach to controlling point-source polluters through a combination of pollution permits and natural resource charges.[1] This mixed system of permits and charges, which is a carryover – or perhaps hangover – from the Soviet period, was substantially revised with the Law on Natural Resources Tax passed in September 1995. To provide some perspective on the Latvian economy, and perhaps to put environmental policy in some context with the economic transition, Section 2 provides a brief overview of major economic changes that have occurred during the transition to a market economy since re-establishment of independence in 1990. Section 3 provides a synthesis of the current mixed system of standards and taxes based on pollution permits and the related system of (newly revised) natural resource taxes. Exam-

---

1   In Latvian, the word 'tax' is used to denote various types of payments, such as taxes to fund the government and charges for consumption of natural resources and discharge of pollutants into the environment. We shall mainly use the word tax in this chapter, for consistency with the Latvian translation of the law, even though some of the payments specified in the law are probably better thought of for this book as charges for consumption of environmental resources.

ples of the resource tax impact on specific enterprises in key sectors are also provided. Section 4 takes a closer look at implementation experience based on reported pollution levels and permit information for the main point-source air polluters in the two largest cities in Latvia – Riga and Daugavpils. Section 5 concludes with recommendations for further improvement in the system of permits and taxes.

2   ENVIRONMENTAL IMPACTS OF THE ECONOMIC TRANSITION

Given the internal logic of the Soviet Union and the new needs and incentives of a market economy, it is not surprising that gross domestic product fell about 50 per cent between 1990 and 1994 (CSBL, 1995e). The economic transition associated with this contraction has had profound, largely beneficial, impacts on the environment through changes in the size and composition of agricultural and industrial output and land use.[2] It is necessary to understand this structural transformation of the Latvian economy to have a perspective on the differences between past, current and likely future environmental problems.[3]

As in other parts of the former Soviet Union, where agricultural production was not based on comparative advantage and logical relative prices for inputs and outputs, Latvian agricultural output has fallen substantially and a number of state farms have dissolved into small private farms.[4] Between 1990 and 1994, the quantity of agricultural production fell about 50 per cent, with about a 30 per cent drop in crop production and a 70 per cent drop in livestock production. Sown area fell from about 1.627 million hectares in 1990 to 1.194 million hectares in 1994, with most land allocated to cereals, pulses and potatoes. As of 1994, about 901,000 hectares were sown by 'private' farms (peasant farms, household plots).

---

2   All agricultural data reported in this section come from CSBL (1995c).

3   For reference, life expectancy at birth in Latvia is estimated to be 60.7 for males and 72.8 for females (CSBL, 1995a) as compared to 61 for males and 63 for females in low-income economies and 66 and 72 for upper-middle-income countries (World Bank, 1994). If the statistics are true, this is a rather striking difference.

Inflation was 109 per cent in 1993 after the introduction of the lat, and 35 per cent in 1994 (CSBL, 1995e).

4   For reference, as of 1 January 1995, there were 95 state farms, 656 statutory companies, 64,000 peasant farms, 118,000 household 'plots', and 124,000 subsidiary farms. 'Peasant' farms are private farms. Between 1990 and 1994, the number of private 'peasant' farms increased from about 7,500 in 1990 with an average size of about 14.5 hectares with 8 hectares sown to about 64,300 farms in 1994 with an average size of about 12 hectares with 6.8 hectares sown.

The sown area in state and collective farms and statutory companies fell from about 1.482 million hectares to about 293,000 hectares.

As part of the breakup of large-scale farming, essentially the deindus- · trialization of agriculture, relative prices of outputs to inputs have fallen with the resulting declines in the use of different inputs (energy, machinery, inorganic chemicals and pesticides). At the same time, however, poorer-quality land was probably taken out of production and labour was probably more efficient and intensively used per hectare. In combination, yields of cereals and pulses fell from about 2.36 tons/hectare in 1990 to about 1.84 tons/hectare in 1994, while potato yields remained at about 13 tons/hectare.[5] Inorganic fertilizer use (mineral content) fell from 363,000 tons to about 17,600 tons in state farms, collective farms and statutory companies, while information on small-scale farms indicates limited inorganic fertilizer use.

The main implication of this agricultural contraction (less land, fertilizers, pesticides and livestock) is reduced nonpoint-source pollution from the agricultural sector. To be sure, there are in some locations continued pollution problems from existing stocks of manure, old chemical storage, and perhaps continued discharges from chemical stocks (phosphorous) built up in soils. But to a very significant degree, for now at least, market changes have substantially reduced agricultural nonpoint-source pollution problems. The future of nonpoint-source agricultural pollution will depend heavily on future agricultural policy, and especially if and how heavily agricultural production will be subsidized in the future.

Industrial production has followed a very similar path to agriculture.[6] For example, industrial output fell about 40 per cent in 1993 and about another 10 per cent in 1994 (CSBL, 1995e). In constant 1993 prices, manufacturing fell from 882 million lats in 1990 to about 300 million in 1993 (SCSL, 1994). Between 1990 and 1993, mining output fell about 60 per cent and public services supply (electricity, thermal energy, gas and water supply) fell about 40 per cent (SCSL, 1994).[7]

5   This 22 per cent decline in cereal yield is equivalent to about 8 years of average yield growth in developed market economies' agricultural sectors.
6   While privatization of agriculture and retailing proceeded quickly, about 66 per cent of industrial output occurred in state enterprises and another 13 per cent occurred in some form with state participation as of 1993 (SCSL, 1994).
7   For reference, peat production fell from 2.8 million tons a year in 1990 to about 0.4 million tons in 1993 (SCSL, 1994).
    Regarding trade, the EU and the newly independent states (mainly Russia) are the main trading partners (CSBL, 1995e). The value of exports during the January–August period of 1995 was about 437 million lats. Of this total, wood and wood products accounted for about 120 million, textiles about 59 million, food products about 41 million, and base metals about 36 million (CSBL, 1995e).

One clear result is that energy and water consumption declined due to the contraction. For example, water consumption declined from 595 million m^3 in 1991 to about 406 million m^3 in 1993 (SCSL, 1994). Electricity production fell from about 6.6 million kWh in 1990 to about 3.9 million kWh in 1993 (SCSL, 1994), and thermal energy fell from 26.1 million Gcal in 1990 to about 13.1 million Gcal in 1993 (SCSL, 1994).[8]

Transportation provides a mixed picture. For example, cargo traffic at Latvian ports fell from about 318 million tons in 1990 to about 59 million tons in 1994 (CSBL, 1995d). Oil transported by oil pipelines increased from 14.9 million tons in 1993 to 15.3 million tons in 1994 (CSBL, 1995d).[9]

Mobile source air emissions from private cars continue to increase. About 251,000 cars were reregistered after 1 January 1993 (CSBL, 1995d). The data on motor vehicles are somewhat difficult to interpret because the data beginning with 1994 include only those cars registered or reregistered after 1993. For example, there were about 282,000 cars registered in 1993 and 252,000 reregistered in 1994. About 30,233 cars were registered for the first time in 1994, which implies about a 10 per cent growth rate in the car stock (CSBL, 1995d). However, unreferenced data in the Latvian National Environmental Policy Plan report that the car stock increased from 283,000 in 1990 to 391,000 in 1993 (MEPRDL, 1995, p. 43). Regardless of the exactly correct number at this moment, it is clear that car use, mobile source emissions from leaded gasoline, and congestion may become a potentially serious urban environmental problem, mainly in Riga, for the future.

3   POINT-SOURCE POLLUTION PERMITS AND CHARGES

Within the substantial structural changes taking place in Latvia, environmental policy has begun to develop and evolve to meet the needs of a modern market economy with an eye towards integration with the European Union. One important component of overall environmental policy is the system of pollution permits and taxes. As mentioned in the

8   Forestry is one area of strong growth that could raise some environmental concerns. Forests cover about 44 per cent of the country, which is about 2.8 million hectares (SCSL, 1994). Forest harvests increased from 3.8 to 4.8 million m^3 between 1990 and 1993, with the amount harvested on state forest lands increasing from about 2.4 to 3.8 million m^3 during the same period (SCSL, 1994). This amount remains considerably less than estimates of sustainable forest harvests. There is also the issue as to whether Latvia is earning sufficient benefits from the harvesting of public forests.

9   There are 426 km of oil pipelines, 329 km of oil products pipelines and 1,237 km of natural gas pipeline (CSBL, 1995e).

introduction, Latvia follows a mixed approach of pollution permits and pollution taxes, which includes self-reporting of pollution levels, state inspection, some ambient and emission standards, certain types of fines for not having permits or not paying charges in time, and potential court action. The purpose of this section is to take a closer look at the logic and structure of this system and new changes under way following passage of the Law on Natural Resources Tax in September of 1995. In the following, except where noted, any reference to the tax law refers to the 1995 Law on Natural Resources Tax.

The basic approach can be summarized as follows. In pollution permits, a limit or threshold level for a pollutant is specified. The limits can include quantity per time period (grams per second, tons per year) and concentrations (milligrams per ton or litre or cubic metre). Since the limits vary across point sources, these limits could be called source-specific performance standards. The details of the permitting process are described in Section 3.1. A base pollution tax is charged for all units of pollution up to the limit, and then an extra rate (four times the base rate) is charged for all units above the limit. The details of the taxes are described in Section 3.2. Given the potentially large jump in the tax rate at the limit, the actual limit specified in the pollution permit is a key policy parameter. The pollution taxes within limits are considered to be costs of production for income tax purposes. Pollution taxes for above-limit amounts are considered to be penalties and are paid out of after-tax income. If pollution occurs by an enterprise without a permit, or the enterprise does not truthfully self-report (and the pollution is discovered) then the enterprise is liable for paying twelve times the base rate for each unit of pollution.

A combination of economic efficiency, common sense and political reality created this mixed system. This structure of pollution taxes is not exactly consistent with the standard formulation of pollution taxes equating marginal pollution damages and abatement costs (for example, see Tietenberg, 1991). However, if all enterprises are polluting within their limits, or above their limits, then the Latvian system essentially works as a pure tax approach that equates marginal abatement costs of all enterprises. If all enterprises are at their limits, then the environmental goal (as specified in the limit) is achieved at the potential expense of cost efficiency because marginal abatement costs are not equated across firms. In fact, marginal abatement costs could vary between one and four times the base pollution rate. Of course, it is an empirical question to know just how 'inefficient' this situation would be.

However, substantial uncertainty exists regarding marginal abatement costs and environmental damages. For example, as Weitzman (1974) showed and as explained in simple words by Tietenberg (1992), a standards approach is preferred to a tax when the marginal environmental damage function is steep compared to the marginal abatement cost function. Thus, if such conditions are true in Latvia for some pollutants such as extremely toxic pollutants, then the combined tax and limit approach is preferred to just a tax. The opposite is also possible in Latvia. In fact, given the lack of knowledge with cost-efficient pollution abatement, it could be argued that a tax approach is preferred on efficiency grounds.

And, of course, history and politics played an important role in shaping the system. The basic structure of permits, limits and a two-tiered tax follows directly from the former Soviet system of environmental policy. While it is well known that pollution permits were not always – or even usually – enforced, Latvia is now in the process of trying to develop, implement, monitor and enforce the system of permits and taxes as a major component of the country's overall policy package. It is using a permit and limit system largely because it existed at independence in 1990. Soviet period environmental standards (ambient and effluent in some cases) remain in force, to the extent that they are not absolved in Latvian law, and there has been little movement to revise the system of standards to be more implementable and perhaps consistent with developments from other parts of the world.

3.1 Obtaining a Pollution Permit

Table 4.1 summarizes key components of the pollution permitting system in Latvia. The table is not intended to be exhaustive, but it highlights major parts of the process, the legal and/or regulatory basis, and which organization in the government is mainly responsible for developing or implementing policy. In the table, parts 1, 2 and 3 are steps in obtaining a permit, while parts 4, 5 and 6 provide some basis for determining limits and monitoring pollution levels.

A main law for pollution policy is the Law on Environmental Protection adopted 6 September 1991. This law mentions environmental quality norms and standards (Section 4, Article 18) to be determined and monitored by Regional Environmental Protection Boards, which were recently renamed Regional Environmental Boards (REBs).[10] It also mentions that

10   The nine Regional Environment Boards in Latvia, which are under the Ministry of Environment Protection and Regional Development, cover the country's 26 administrative districts. These REBs are responsible for implementing and to some degree enforcing the system of permits and taxes.

The law includes references to liability related to environmental protection, compliance with environmental quality standards, and complying with environmental laws in general. There is also a clause on 'mandatory environmental insurance' which has not been implemented to date.

issuing pollution permits should be carried out by the regional boards (Section 5, Article 25). The REBs can revoke permits as well. A related regulation 'On the Republic of Latvia State Environmental Inspection' (from 10 October 1990) identifies that a permit is needed for emission of polluting substances into the environment. In practice, this has included permits for air emissions, water use and effluent, and solid waste.

The basic characteristics of the permitting system can be summarized in general, while the details vary depending on the nature of the pollutant (air, water, solid waste, and so on). For the purposes of this chapter, we shall focus mainly on air emission and water effluent permits. To start the process, enterprises are responsible for filling out an application and providing technical data on the enterprise and likely pollution levels. The enterprise delivers the application and supporting materials to the REB. In the application, the enterprise makes a request for various permitted levels of pollution.[11]

To obtain a water permit, the application states the need for the water, describes the type of water use, the source and volume of water to be consumed, as well as, for direct dischargers, technological capacities and conditions of wastewater treatment. These conditions include the volume of wastewater discharge, the quality and the expected concentration of pollutants in the wastewater.

When reviewing an application for a water permit, the REB is supposed to follow existing ambient and effluent standards specified in the 'Water Usage Resolution No. 3 of November 1991' and the government regulation 'Water Pollution Hazard Categories and Their Maximum Allowable Concentrations in Water Bodies and Wastewater Discharge Sites'.[12] Information on existing ambient water and effluent standards are provided in Table 4.2, and for ambient air in Table 4.3. These are former Soviet standards.

The ambient standards probably need to be revised. For example, for air, the 20-minute sample average is probably too short for short-term exposure and there is the need to take into account the probabilistic nature of ambient air quality. A 95 or 98 per cent compliance rate with some maximum concentration is probably preferred to the current situation. Such changes do not automatically imply a 'loosening' of ambient standards; they do, however, imply a redefinition of ambient standards.

---

11 The application letters should be approved by the local (county) municipality.
12 This regulation is a 'Statute of the Ministry of Environmental Protection and Regional Development' accepted by the Regulations of the Cabinet of Ministers No. 57 (21 March 1995). With regard to water protection, the water code that was accepted in 1973 by the Latvian SSR Supreme Soviet is currently in force.

Table 4.1   A summary of the pollution permit system in Latvia

Policy components	Legal basis	Responsible institution
1. Need for permit	Law on Environmental Protection (Article 25, Sect. 5) Law on Air Protection (Article 16, Sect. 5) Regulation on State Inspection of Environmental Protection of LR (10 Oct. 90)	State Env. Inspection (SEI) Regional Environmental Boards (REBs)
2. Structure permit	Water Usage Resolution No. 3 of Nov. 1991. Instruction on temporary air pollution permits for stationary pollution sources of VARAM, 24 Jan. 1995. Permit form No. 1	REB
3. Permit decision	Regulation On State Environmental Inspection of the LR (1990, Sect. 2)	Inspectors of REB
4. Ambient standards		
–air	USSR Normatives on Max. Allowable Concentration of Air Pollutants, No. 10–6 22/–853 from 14 Sept. 1995.	REB SEI EIA Dept
–water	Water Pollution Hazard Categories and their Maximum Allowable Concentrations in Water Bodies and Waste Water Discharge Sites, adopted by the regulations based on the Statute of the VARAM 12 March 1995)	
5. Emission/effluent standards		VARAM
–air	No standards	
–water	Same as for Ambient Water	
6. Monitoring	Law on Air Protection (Article 18, 1982) Water Usage Resolution No. 3 of 1991 (Article 5)	Enterprise or permit owner REB

The legal basis for an air permit, besides the Environmental Protection Act, is also specified in the Law on Atmosphere Protection from 1982 adopted by the LSSR Supreme Council. This law states that every enterprise which emits polluting substances or has projected, reconstructed or existing sources of pollution, must have a permit for emission of polluting substances into the atmosphere (Section 5, Article 16).

To apply for an air permit, enterprises are supposed to supply relevant information on the production structure, pollutants and pollution control technology.[13] The enterprise submits these data to the REB. For enterprises of state importance, the application has to be approved first by the Environmental Impact Assessment Department (EIAD) of VARAM (the Environmental Policy Protection Plan of Latvia). If approved by the necessary institutions (EIAD and/or REB), the REB issues the permit.

Upon receiving a complete application package, the REB has one month to make a decision. Two types of permits are given: a temporary permit, which is valid for one year; and regular permits issued for up to five years depending on the agreement and situation.[14]

Both parties must agree on what pollutant discharge level is acceptable, and then the permit owner pays taxes quarterly based on its own calculations. The REBs in principle verify the data submitted through direct sampling and laboratory analyses or through indirect material/flow balances. REBs are also responsible for monitoring the technology utilized and doing laboratory analyses of water/air according to an inspection schedule.

It is not really clear how well pollution levels are determined and reported by enterprises, and how REBs are able to monitor and enforce the system. As will be seen in Section 4, enterprises do have permits and do report pollution levels. However, it is not clear in many situations how the pollution levels are monitored/estimated, and if enterprises report truthfully. It can look suspicious if reported pollution levels equal permitted limits (down to decimal places), which occurs in some cases. There is currently little on-line monitoring technology. It is not clear if the scheduled enterprise and REB monitoring, along with some unannounced sampling, is sufficient. Further analysis is needed to determine how to

13   Specifically, this information includes: characteristics of production; number of sources of polluting substances; work schedule for emission sources; location of sources; quantity of polluting substances for each source; characteristics gases such as volume (cubic metre per second), temperature; emissions of polluting substances (name, maximum emission rate in grams per second, and tons per year); any plans for reducing emissions and characteristics of existing control technology.

14   Temporary air permits are issued according to the instruction of VARAM 'On the Procedure of Issuance of Temporary Permits for Stationary Pollution Sources' (dated 24 January 1995).

Table 4.2   Water pollution hazard categories, ambient standards and effluent standards from Latvian list of 36 pollutants[a]

Pollutant	Hazard category	Ambient (mg/l)	Effluent (mg/l)
1. Suspended solids	1	–	15
2. Chloride	1	300	300
3. Sulphate	1	100	500
4. BOD(21)	2	3	12
5. COD	2	15	90
6. Nitrate	2	9.1	25
7. Nitrite	2	0.02	0.15
9. Phosphate	2	0.25	2.5
10. Detergents (anionic)	2	0.1	0.5
11. Detergents (non-eugenic)	2	0.1	2.0
18. Oil products	3	0.05	0.5
19. Phenol	3	0.001	0.02
22. Formaldehyde	3	0.05	0.5*
23. Sulphite	3	1.9	5*
32. Cadmium	4	0.001	0.02*
36. Lead	4	0.03	0.1*

Notes:
a.   Numbers before pollutant name correspond to number in list of 36.
     In the last column, * denotes a zero level allowed for discharges to small rivers, lakes and ditches. Maximum allowable concentrations for other pollutants not in the Latvia national list are taken if necessary from the former USSR normatives and standards.

improve cost effectively the information, monitoring and enforcement requirements of this permitting system regardless of any revisions in the general system of standards.

3.2 The Structure of Pollution Taxes

Pollution taxes have been part of Latvian environmental policy since shortly after independence in 1990. Due to several factors, including a generally low level of taxes combined with erosive effects of inflation, the 1991 Law on Natural Resources Tax has had little impact on pollution-control incentives and has generated little revenue.

During 1995, two main policy developments have formalized the political will and legal structure – at least on paper – to develop, implement and

Table 4.3 *USSR/Latvian standards on maximum allowable concentration of air pollutants for populated areas (No. 10–6 22/–853 as of 14 September 1990*

Pollutant	Maximum allowable concentrations of air pollutants*	
	MAC mg/m³ (20 minutes)	MAC mg/m³ (24 hours)
SO_2	0.5	0.05
Particulates	0.5	0.15
NO_2	0.085	0.04
CO	5	3

Notes: *These concentrations are for general environment, there exists also separate normatives in maximum allowable concentrations for work places. These four air pollutants account for more than 80 per cent of reported air pollution in permits. There are an additional 160 pollutants regulated in principle.

enforce various types of 'economic instruments' as part of Latvian policy. First, in May 1995 the Cabinet of Ministers (the government) accepted a general guidance document developed by VARAM, which recommended several basic concepts, such as the polluter pays principle, and the use of economic incentives for environmental protection.

Second, a more recent development is the passing of the general Law on Natural Resources Tax in September of 1995. In brief, this law allows the possibility for effluent taxes based on pollutant volume, product taxes based on weight or value, credits from such payments to subsidize environmental investments, tradable pollution licences based on limits in existing permits, and some form of subsidies for the collection, reuse or appropriate disposal of various materials.

It is safe to say that the process of selecting taxable items and tax rates in the new law was based on political acceptability. There was no hint of a Pigouvian notion of taxation (that is, marginal pollution damages equal to marginal abatement costs) in the debate in the Latvian parliament (the *Saeima*) during the three rounds of voting which took place over several months. At this stage, it is probably best and safest to interpret the new law as having created a more organized and flexible foundation for environmental taxes in Latvia. Several Cabinet regulations will have to be developed, approved, implemented and enforced before any substantial impact on pollution control can be expected.

Given the recent legal change, we shall focus our discussion mainly on the structure of the new tax law, making specific reference to the old law where relevant. Regarding the new tax law (the 1995 Law on Natural

Resources Tax), it is first necessary to be clear on the commodities that are actually taxed.

For this tax law, the term 'use of natural resources' is used to denote four broad groups of taxable activities. These groups are: (i) discharge of pollutants directly into the environment (see Table 4.4); (ii) direct extraction of certain natural resources (see Table 4.5); (iii) use or consumption of certain products defined to be harmful for the environment (see Table 4.6); and (iv) any kind of activity or absence of activity degrading the environment or natural resources.

Table 4.4 shows that group (i) 'use of natural resources' are direct charges on the quantity of air emissions, water effluent, solid waste disposal at disposal sites, and solid waste generation and disposal in other areas. On average, one impact of the new law is about a 300 per cent increase in tax rates in one year, at a time when inflation is about 25 per cent. This change does imply a real increase in the tax.

While it was mentioned earlier that the parliament did not directly consider the damage side of the issue when choosing tax rates, Table 4.4 shows that the general structure of direct emission taxes is based on general categories of pollutant 'hazard'. A Cabinet of Ministers regulation is needed to formalize the list of pollutants that fall into each category. It is possible to inject some more formal logic into the final allocation of pollutants to each category to improve the relative nature of the direct pollution taxes.[15]

In general, group (ii) is designed to cover traditional mining and water extraction activities, but does not cover Latvian forests. As reported in Table 4.5, rates for surface and underground water use increased about 100 per cent between 1995 and 1996.

Table 4.6 reports the list of group (iii) direct product charges ('input charges') that are considered to be indirectly linked to pollution and environmental damage. From Table 4.6, these products include lubrication oils, batteries and accumulators, ozone-depleting substances, mercury lamps, tyres, glass, plastic and metal containers for foods and drinks, disposable plastic dishes and

15   To implement the new system of taxes, some government regulations are needed to specify which pollutants fall into the different categories. VARAM intends to harmonize classification of chemicals and polluting substances with standards of the EU. For reference, some examples from the current classification in 1995 are listed here. For air emissions, categories include: non-toxic – dust; medium dangerous – carbon oxides, inorganic compounds of phosphorous and nitrogen; dangerous – oxides of sulphur and nitrogen, ammonia, vapours of organic compounds; highly dangerous – lead, mercury, vanadium. For water effluents, categories include: non-toxic – sulphates, alkaline chlorides; medium dangerous substances – phosphorous, nitrogen (nitrites, nitrates), BOD; dangerous substances – oil and oil products, phenols, aromatic hydrocarbons; highly dangerous substances – cyanides, heavy metals, pesticides.

Table 4.4   *Examples of pollution taxes in Latvia (base rates within limits)*

	1996 tax	1995 tax
Air emissions		
Non-toxic (lats/ton)	3.00	1.00
Medium dangerous (lats/ton)	4.50	1.50
Dangerous (lats/ton)	10.00	3.25
Highly dangerous (lats/ton)	800.00	3,250.00
Water effluent		
Non-toxic (lats/ton)	3.00	1.00
Suspended solids (lats/ton) (non-toxic)	10.00	1.00
Medium dangerous (lats/ton)	30.00	10.00
Dangerous (lats/ton)	8,000.00	2,500.00
Highly dangerous (lats/ton)	50,000.00	15,000.00
Waste storage (disposal)		
Non-toxic waste (lats/m^3)	0.25	0.15
Toxic waste (lats/m^3)	1.50	0.50
Highly toxic waste (lats/m^3)	50.00	15.00
Tax rates for pollution of soil, ground and water beds		
Medium dangerous (lats/ton)	100.00	–
Dangerous (lats/ton)	1,000.00	–
Highly dangerous (lats/ton)	10,000.00	–

tableware. Some of these charges are based on weight, some on 'units', and some on value before customs duties or value added tax.

The main objective of developing these taxes is to discourage consumption and increase reuse/recycling and appropriate disposal of these products in Latvia. Some fixed amount of tax payments for these product charges are earmarked for subsidies for collection, reuse, recycling and appropriate disposal. In principle, this could be considered some form of a deposit-refund scheme, where the deposit (the tax) is not totally returned to the depositor. To implement this part of the law, it is necessary for the Cabinet of Ministers to develop regulations that specify the process and procedures for collecting the tax and the potential refund/subsidy process. While earlier drafts of the natural resources tax law included excise taxes on the consumption and import of energy resources, this section was eliminated by the parliament during the second reading of the tax law.

Table 4.5   Water extraction/use/consumption taxes in Latvia (base rates within limits)

Water use	1996 tax	1995 tax
Underground water (lats/m^3)	0.01	0.005
Surface water (lats/m^3)	0.002	0.001

It is expected that the Ministry of Environmental Protection and Regional Development will return to such a proposal in the near future. It would be useful to consider some form of general equilibrium analysis of the Latvian economy to understand the impacts of such energy taxes before they are debated as policy.

Group (iv) is a 'safety-net' category that shows up in Latvian laws (for example, the Law on Environmental Protection) that is not usually defined but that seems to be included for liability reasons. In the Law on

Table 4.6   New product charges (as of 1996)

Item	Charge
Mineral (lubrication) oils	0.02 lats/l
Accumulators and batteries	
–lead <50ah	1.50 lats/unit
51–100ah	3.00 lats/unit
101–150ah	4.50 lats/unit
>150ah	6.00 lats/unit
–other	15% of item value
Ozone-depleting substances	(ODR*1 lat)/kg[a]
Mercury luminescent bulbs	0.10 lats/unit
Tires 0.05 lats/kg	
Packaging (food and drinks)	
–glass 3% of item value	
–plastic pack	5% of item value
–metal containers	5% of item value
–cardboard pack	4% of item value
Disposable plastic dishes and tableware	50% of item value

Note: a. ODP equals the ozone-depletion factor (an index) for a specific commodity.

Environmental Protection, there seems to be a negligence–based approach to environmental liability written into the law, where it is necessary to show that an action violated some law or related regulations before being held liable for certain payments or even cleanup requirements. In many circumstances, group (iv) could help to define some action as violating a negligence requirement. In specific circumstances, the actual taxes in groups (i)–(iii) would be used to define the rates for this type of 'use of natural resources'.

Tables 4.7 and 4.8 provide some indication of the overall likely magnitude of basic air and water taxes based on information on 1994 levels of pollution and the new rates specified in the law. Table 4.7 shows that total SO_2 charges would be about $1.5 million and NO_x taxes would be about $0.30 million in Latvia, based on 1994 pollution levels and assuming the whole amount was assessed at the basic rate in the tax law. It will be shown in Section 4 that this is a pretty good assumption at the moment in Latvia. If all of these taxes were actually collected, which is highly unlikely, and passed on to domestic consumers, this would imply a per-capita tax payment of about $0.46 for NO_x and SO_2. Not surprisingly, the overall payment falls on the main urban areas of Riga and Daugavpils.

Given aggregated information for Latvia on BOD_7 discharges, Table 4.8 reports implications of the new law based on hypothesized discharge concentration based on currently reported categories in national statistics (that is, whether discharge is within the norm/limit, above the norm but treated to some degree, and above the norm and not treated at all). Based on the effluent quality assumptions in the table, about $1.5 million would be paid in BOD_7 taxes if the tax was collected, yielding a $0.63 per-capita tax.

Some simple examples of the natural resources tax burden on specific, large enterprises in various industries were developed by the authors to consider the economic impact of the new law. The results of these calculations are based on reported pollution levels multiplied by the new tax rates, summarized in Table 4.9.

For enterprises in food-processing industries (bread, meat and milk), the magnitude and structure of the tax varies considerably. The bread industry pays the least, about 1,798 lats (about $3,400), the majority of which is for water effluent. The meat enterprise pays primarily for water consumption. The potential importance of the product charges in the tax law become apparent for the milk enterprise, whose tax payment is calculated at more than 60,000 lats, primarily in higher packaging costs. None the less, the estimated tax payment relative to revenues looks rather small (less than one per cent).

Of course, this information should be interpreted with caution. It is possible that profits for some enterprises are quite small relative to

Table 4.7 General implications of Latvian natural resources tax system with 1994 reported air pollution levels

Variable unit Location	SO_2 (tons)	Tax ($m.)	NO_x (tons)	Tax ($m.)	Pop^a (m.)	SO_2/p^b (tons/p)	NO_x/p^c (tons/p)	tax/p^d ($/p)
Latvia total	51,598	0.980	10,281	0.195	2.566	0.020	0.004	0.458
Riga region	15,838	0.301	3,358	0.064	0.147	0.108	0.023	2.481
Riga city	8,000	0.152	1,683	0.032	0.856	0.009	0.002	0.215
Liepaja	2,785	0.053	1,579	0.030	0.105	0.027	0.015	0.792
Daugavpils	5,844	0.111	574	0.011	0.121	0.048	0.005	1.006
Rezekne	2,828	0.054	156	0.003	0.042	0.067	0.004	1.339

Notes:
a. Pop is Latvian population for 1994.
b. SO_2/p is SO_2 per capita.
c. NO_x/p is NO_x per capita.
d. Tax/p is tax in dollars per capita.

Table 4.8   General implications of Latvian natural resources tax system with 1994 reported water discharges and possible effluent concentration quality

	Cubic metres discharge (000 m_3)	BOD (g/m_3)	Tons (BOD$_7$)	Tax ($m.)	Tax/person ($/person)
Clean	119,336	0	–	–	–
<Norm	66,721	30	2,002	0.114	–
>Norm, WT	176,048	100	17,605	1.003	–
>Norm, NT	39,769	200	7,954	0.454	–
Total	401,874	–	25,561	1.571	0.61

Note: BOD$_7$ = $57/ton.

revenues, in which case the tax burden relative to profits (revenues minus production costs, other taxes, and so on) could be quite high. Good information on costs is not available for these enterprises.

The last three enterprises in Table 4.9 are municipal service enterprises, where revenues are based on tariff schedules. In this case, water effluent and solid waste disposal charges are high relative to other large enterprises and high relative to revenues. Given the monopoly position of these two enterprises, these taxes will be passed on to consumers, which could provide some additional incentives for conservation depending on the relationship between, for example, water consumption, sewerage and service charge.

The tax law includes the possibility for the use of other environmental policy instruments, such as tradable pollution rights and natural resource tax credits for environmental investments. Regarding tradable rights, the plan is to modify the current system by redefining the quantitative limits in pollution permits as a quantity of tradable 'licences' owned by the firm (with the same time period as defined in the permit). It is unclear exactly how this exchange between permit limits and transferable licences will take place for existing permits. For new licences that may be issued, the tax law states that the licences will be sold for the price of the tax on the item. If a system of tradable licences is developed, the total natural resource use specified in permits and licences will be considered the limit for tax purposes. Section 4 discusses the issue of tradable licences further.

A last component of the tax law of interest for this discussion is the possible use of specific tax credits. Such credits can be granted under the

Table 4.9   Natural resources and environmental tax burden (lats) for large enterprises

Enterprise	Total tax	% of rev. of tax	Main component
Bread	1,798	0.06	Water effluent
Meat	27,161	0.15	Water consumption
Milk	63,443	0.68	Packaging
Building construction	2,010	0.80	Mineral consumption, air emissions
Gypsum products	9,040	2.56	Mineral consumption
Bricks	1,490	0.47	Mineral consumption
Paper	754,420	0.12	Water effluent, air emissions
District heat	4,761	0.07	Air emissions
Water and sewerage	3,547,197	33.57	Water effluent
Solid waste disposal	320,781	28.01	Water effluent, solid waste disposal

Notes: These calculations assume the new tax rates, existing permits, and predicted 1995 pollution and output levels. Complete data are available from Janis Brunenieks.

Law on Natural Resources Tax to enterprises that finance projects aimed at some 'reduction' of environmental pollution. Currently, three general ideas are contained in the law. First, the maximum amount of a tax allowance is supposed to equal the tax value of the pollution reduction due to implementing the pollution-control activity (referred to as the 'project' in the law). It is not clear in the law, and it will need to be made more clear through future regulations, from what level the reduction will be calculated (past levels, projected future levels).[16]

The second idea is that the credit is essentially an interest-free loan with a conditional grace period equal to the agreed-upon length of a

16 All the revenues from these taxes are earmarked to various special budgets at the national, district and county level. In this sense, the environmental taxes are a 'mixed' tax, falling within the jurisdiction of three levels of government. The administration and enforcement requirements fall on the Ministry of Environmental Protection and Regional Development and the Ministry of Finance at the national level. However, the special budgets are the responsibility of different levels of government. As a result, it could be necessary for an enterprise to get a credit from three different levels of government to make complete use of this tax credit provision.

project implementation, which could be multiple years in some cases. At the agreed-upon completion date, if the project is fully implemented and continues in operation, then the interest-free loan can be converted into a direct grant to the enterprise. If the project is not adequately implemented, then the repayment on the loan begins with interest from the origination date.

The third idea is that enterprises will have to co-finance the activity to some degree. Thus, in general, enterprises will select projects where the private benefits of the investment to the enterprise at least cover its direct share of the costs of the project. The remaining share is financed through the pollution tax credit, which pays for the social benefits of the project.[17]

4   FIRM-LEVEL AIR EMISSION DATA FOR RIGA AND DAUGAVPILS

While the previous section outlined the general structure of permits and the recently passed tax law, this section takes a closer look at a sample of enterprises with air pollution permits from Riga and Daugavpils. Riga is the capital and the major metropolitan centre of Latvia with about 40 per cent of the population, while Daugavpils in the southeast along the Daugava river is the second largest city in terms of population.[18]

The data reported in this section are based on enterprise-specific permit information provided by the Latvian Environmental Data Centre, which is an institution under the general responsibility of the Minister of Environmental Protection and Regional Development. The data set includes specific information on pollution permits for both cities covering the years 1992, 1993 and 1994. To focus this discussion, we concentrate on two point-source air pollutants – sulphur dioxide and nitrogen oxide.

Table 4.10 summarizes basic information on reported pollution levels and permitted amounts for the complete sample for all years. The last column shows such information for the 20 largest SO_2 polluters in 1994.

---

17   To receive a tax allowance, an enterprise cannot have any tax debts on any type of tax (income, value-added, work social taxes, and so on), which will greatly restrict the possibility of receiving a tax credit for many enterprises.

18   Enterprises covered in the data set for each city vary across the years because of changes in business structures (bankruptcies, mergers, name changes, and so on). For reference, the number of enterprises in 1992, 1993 and 1994 for Riga are 78, 65 and 68 and for Daugavpils are 37, 39 and 35. In the data set, all types of dust and ash are aggregate and referenced as 'particulates'. Given the aggregation across types, no permitted levels are reported.

Table 4.10   Total SO$_2$ and NO$_x$ in Riga and Daugavpils

Year	1992	1993	1994	1994 (20 largest SO$_2$)
Riga				
1. SO$_2$ Actual	5,040	6,463	7,825	7,551
Permitted	11,366	12,976	15,169	13,984
2. NO$_x$ Actual	1,822	1,415	1,379	952
Permitted	3,242	3,059	2,872	2,013
3. Particulates Actual	1,483	902	969	458
Daugavpils				
1. SO$_2$ Actual	4,869	5,584	5,843	5,843
Permitted	8,085	7,638	8,681	8,681
2. NO$_x$ Actual	594	582	570	570
Permitted	879	887	841	841
3. Particulates Actual	625	457	343	321

From Table 4.10, several points stand out. For both cities, reported levels of SO$_2$ emissions increased, which is consistent with observed fuel switching away from natural gas, where prices have risen substantially, to lower-quality heavy fuel oils. As compared to the complete data set, the 20 largest SO$_2$ polluters in 1994 contain the vast majority of reported and permitted SO$_2$ and NO$_x$ emissions. Thus, just 20 enterprises in both cities provide a pretty good picture of aggregate point-source polluters.

The data also show that, in aggregate, permitted emission levels for SO$_2$ and NO$_x$ are almost 100 per cent larger than actual emission levels in Riga for all years, while the difference is somewhat smaller in Daugavpils. Three reasons explain this difference between actual and permitted emissions. First, given the probable poor quality of monitoring technology, reported emissions may be a poor indicator of actual emissions and enterprises could have an incentive to err on the safe side and under-report emissions. Second, given the substantial declines in industrial production and energy demand in Latvia, it could be that production levels have just declined, with the resulting declines in emissions, while permitted limits have remained constant. And third, the permitting process is lenient in some cases and enterprises are able to receive permits with large limits. It is likely that an answer lies in some combination of all three.

Given the importance of limits in permits for the new tax law, Table 4.11 reports the number of enterprises that reported pollution levels above permitted amounts for SO_2 and NO_x. In sum, compliance rates seem exceedingly good based on the reported pollution levels. All enterprises in Daugavpils and Riga were in compliance with NO_x limits in 1994. Also for 1994, all enterprises in Daugavpils were within SO_2 limits, while three of 68 enterprises were above limits in Riga. The last column in Table 4.11 also reports how compliance would change if limits in individual permits were reduced by 25 per cent given reported 1994 emissions levels. For both cities, about 15 per cent of enterprise would become noncompliant. Thus, while existing limits in permits are not necessarily binding on enterprises, such limits could become much more important if an economic recovery lifts pollution levels in the future.

To take a close look at compliance rates if limits changed, Table 4.12 reports how compliance rates would change for the 20 largest SO_2 polluters in each city as limits in permits fell from 1994 levels. For both cities, reducing limits by 10 or 25 per cent for all sources would have modest impacts on compliance rates for the larger polluters. What is perhaps most surprising is that close to half of all polluters would still be in compliance if their limits were reduced by 50 per cent.

The relationship between pollution levels and permitted limits probably has substantial implications for the use and usefulness of developing transferable pollution licences (tradable quotas) based on the new natural resources tax law (Articles 12–15). For example, in 1994, essentially all the polluters were within their limits. This means that, while every enter-

Table 4.11   Number of enterprises with reported emissions above limits

	1992	1993	1994	1994 (25% limit reduction)
Riga				
NO_x	5	3	0	1
SO_2	13	6	3	11
Sample size	78	65	68	68
Daugavpils				
NO_x	0	2	0	6
SO_2	0	2	0	4
Sample size	37	39	35	35

Table 4.12   Noncompliance among 20 largest SO_2 polluters as limits fall

	1994 limit	10% fall	25% fall	50% fall
Riga				
NO_x	0	2	4	7
SO_2	1	1	4	12
Daugavpils				
NO_x	0	3	6	11
SO_2	0	2	4	11

prise but one could be willing to sell a unit of SO_2 or NO_x to some other polluter, there would be no buyers from Riga or Daugavpils.[19]

Of course, the data reported in this section do not consider enterprises in other regions of Latvia. At the national level, there may be more polluters at or over their limits, which helps to set up the right conditions for the use of tradable pollution licences to reduce total costs of pollution control. But the data here provide some caution to environmental policy makers in Latvia to consider carefully the implications of restricting the geographic level of trading.

While the tax law allows for the possibility of tradable pollution licences, it seems unlikely in the short term that SO_2 or NO_x trading possibilities in either Riga and Daugavpils would really have any major benefit at existing pollution levels, limits and tax rates. Limits are high relative to pollution levels and the tax rates above limits are still probably rather low compared to abatement cost options. Since there are no major sanctions imposed on an enterprise which pollutes above limits except for paying the higher tax rate, the cost of being above the limit to an enterprise is relatively low compared to abatement cost options. In fact, if transaction costs were included in the calculation, it might be a net cost to society of trying to develop and implement transferable pollution rights at this time in Latvia. Of course, the data reported in this section are just for two cities, but they are cities that account for more than 40 per cent of the country's population.

---

19   Also note that, when a trading system is in place, SO_2 and NO_x will trade at between $0 and $57 a ton. The low price of $0 reflects the fact that the buyer of the ton of SO_2, for example, will have to pay the base rate tax of $19 even after it is purchased, while the higher price of $57 reflects the above limit price of $76 minus the $19 that enterprise will have to pay as a tax even after it purchases the ton.

It is possible that pollution trading could make immediate sense for some enterprises in the short term. A more complete national analysis is needed of all regions in the country for a larger number of pollutants to determine for which enterprises and pollutants the possibility of transferable pollution rights may be useful. It should also be emphasized here that the usefulness of trading possibilities will probably grow in the future as limits in permits are revised over time and as output levels increase.

Given reported levels of emissions and permitted limits, it is possible to calculate the likely tax implication of various emissions levels under the new natural resources tax law. Taking a look at just the largest 20 SO_2 polluters, Table 4.13 provides an estimate of the tax consequences of the 1994 reported emissions levels. Based on existing hazard categories, NO_x and SO_2 both fall in hazard category 3 with a rate of about $19 per ton (10 lats).

From Table 4.13, SO_2 emissions would account for the majority of taxes for the top 20 SO_2 polluters, which perhaps is not unexpected since the data set draws on the top SO_2 polluters. However, the top SO_2 polluters are largely the top NO_x polluters as well with a couple of minor exceptions. These taxes imply an average annual total tax payment of about $8,000 in Riga and about $6,000 in Daugavpils. It is difficult to judge how 'large' these average tax payments are for specific enterprises, because data is not yet available to compare these tax levels with other costs and revenues.

Another way to look at the numbers is in terms of the investment tax credit clause in the new natural resources tax law (Article 16) discussed in Section 3. As a very simple example, consider an enterprise with a tax bill of $8,000 for about 421 tons of SO_2 (close to the average tax from Table 4.13 for Riga). If some investment or other change was proposed that would 'reduce' pollution 25 per cent from the previous year's level, the enterprise could be able to receive a $2,000 credit from its tax bill to

Table 4.13 *Total tax payments ($) for 20 largest SO_2 polluters under the new natural resources tax law (1994 emissions)*[a]

	NO_x	SO_2	Particulates	Total
Riga	18,103	143,478	2,636	164,219
Daugavpils	10,804	111,016	1,849	123,706

Note: a. It is assumed that both pollutants are in hazard class 3 in the tax law, which implies a tax of $19 per ton for levels within limits and $4 \times \$19 = \$76$ per ton above limits.

subsidize such pollution-control activities. This is a relatively small amount of money.

While total tax liabilities and averages can be determined from Table 4.13, Tables 4.14 and 4.15 report tax payments of the largest SO_2 polluters given their 1994 reported pollution levels and their permitted limit. Both tables also show how tax liabilities would change as limits in permits fell by 10, 25 and 50 per cent. Under the 1994 limits, Tables 4.14 and 4.15 show that tax payments are highly skewed towards a few larger polluters. For example, while the average for Riga for SO_2 is about $7,000, with five of 20 enterprises at or above the average, the median tax payment is about $2,000.

While total tax payments and individual tax payments for SO_2 and NO_x change little if limits were reduced by 10 to 25 per cent for all polluters, some important changes would occur with a 50 per cent decline in SO_2 limits. For example, consider the four big polluters from Riga (r1, r2, r3 and r4) in Table 4.14. With a 50 per cent limit reduction and 1994 pollution levels, taxes for enterprise r1 increase by more than $35,000 while taxes for enterprise r2 increase by only $500. The reason is that the pollution limit for r2 is 800 tons higher than for r1.

Table 4.14 SO_2 tax liabilities ($) for largest SO_2 polluters in 1994 as limits fall

City	Tons (R)	Tons (L)	Tax 1994	Tax (90%L)	Tax (75%L)	Tax (50%L)
r1	1,781	2,314.00	33,839.49	33,839.49	36,434.48	69,408.98
r2	1,618.8	3,220.00	30,757.20	30,757.20	30,757.20	31,258.80
r3	1,188.6	1,638.77	22,583.61	22,583.61	22,583.61	43,629.49
r4	1,036.5	1,620.00	19,693.42	19,693.42	19,693.42	32,603.70
r5	500.9	1,300.00	9,517.18	9,517.18	9,517.18	9,517.18
r6	241.9	549.66	4,595.57	4,595.57	4,595.57	4,595.57
r7	239	448.23	4,542.14	4,542.14	4,542.14	5,394.01
r8	153.3	705.60	2,913.65	2,913.65	2,913.65	2,913.65
r9	132.6	153.42	2,519.48	2,519.48	3,519.33	5,705.52
r10	120	724.53	2,282.58	2,282.58	2,282.58	2,282.58

Notes: Tons(R) is reported pollution in 1994; Tons(L) is permit level; tax (1994) is tax liability with actual pollution level, limit equal to L, and SO_2 considered a category 3 dangerous emission with rate of $19 per ton; and tax (i%L) is the same tax calculation as the permitted level falls to 90, 75 and 50 per cent of the 1994 permitted limit; and r1–r10 are ten specific enterprises.

*Table 4.15   NO$_x$ tax liabilities ($) for largest SO$_2$ polluters in 1994 as
limits fall*

City	Tons (R)	Tons (L)	Tax (L)	Tax (90%L)	Tax (75%L)	Tax (50%L)
r1	111.97	180.00	2,127.41	2,127.41	2,127.41	3,379.64
r2	586.90	1,437.00	11,151.10	11,151.10	11,151.10	11,151.10
r3	59.73	61.63	1,134.87	1,377.66	1,904.63	2,782.91
r4	56.23	59.00	1,068.37	1,246.78	1,751.23	2,591.98
r5	29.99	50.00	569.89	569.89	569.89	854.54

Notes: Tons(R) is reported pollution in 1994; Tons(L) is permit level; tax (1994) is tax
liability with actual pollution level, limit equal to L, and SO$_2$ considered a category 3
dangerous emission with rate of $19 per ton; and tax (i%L) is the same tax calculation as
the permitted limit falls to 90, 75 and 50 per cent of the 1994 permitted limit; and r1–r5 are
five specific enterprises.

5 SUMMARY AND RECOMMENDATIONS

Pollution permits and taxes are a central part of Latvian environmental
policy, which continues to develop and evolve to meet the needs of a
market-oriented economy of Europe. This chapter focused primarily on
the legal and regulatory framework for such permits and taxes, recent
experience with implementation focused on air pollutants in Riga and
Daugavpils, and the implications of this experience for implementing new
provisions in the 1995 Law on Natural Resources Tax.

While existing Latvian policy provides a good foundation for achieving
environmental protection goals, some developments and changes will
help to increase the effectiveness of the current system. Six areas for
development are:

1. As has occurred in other parts of the world (Hahn, 1989), Latvia
    could benefit from separating the discussion of environmental qual-
    ity goals (revisions of ambient goals for specific pollutants) from
    instruments (permits, taxes, environmental impact assessment, and
    so on) for achieving these goals. Identifying goals and revising exist-
    ing ambient standards does not imply 'loosening' environmental
    policy goals; it does mean developing goals that can be monitored
    and are consistent with Latvia's intention of future integration with
    the European Union. Once the environmental goals are clear, policy

instruments can be evaluated in terms of achieving the environmental goal at least cost to Latvia.

2. Regarding the basic system of permits, it would help to clarify the
    legal foundation for requiring permits, formalize the logic for choos-
    ing pollution limits in relation to environmental quality goals, and
    clarify the responsibilities of various levels of government regarding
    issuing, monitoring pollution levels (ambient and emission/effluent
    levels) and enforcing permit requirements.

3. Regarding monitoring and enforcement of effluent/emission levels,
    it is not at all clear that enterprise-level reported pollution amounts
    reflect reality. There is little on-line monitoring equipment at en-
    terprises. The methods used to estimate pollution levels by enter-
    prises, if used at all, may be rather out of date. At a minimum,
    there needs to be an increased focus on the quality of monitor-
    ing/estimating enterprise-level emissions, perhaps through required
    on-line monitoring equipment for some enterprises. It is also clear
    that inspection capabilities/resources of VARAM and the REBs
    need to increase for accurate enterprise-level self-reporting of pol-
    lution levels.

4. VARAM needs to be ready to adjust the Law on Natural Re-
    sources Tax and/or related regulations as the law is implemented
    in 1996. Specific areas of likely improvements could be on revising
    tax levels more in line with related environmental damages and
    abatement costs and procedures for allowing tax credits for envi-
    ronmental investments. At the same time, because of monitoring
    problems or costs, it may be desirable to swap certain pollution
    taxes for related product charges (for example, fuel charges). For
    example, mobile sources are probably the biggest source of NO_x
    emissions in Riga, but they are not covered by the Law on Natural
    Resources Tax.

5. With the passage of the new tax law, it would be very helpful for
    public awareness to improve information management so that infor-
    mation on polluting activities, permits and tax payments is more
    easily available to the general public.

6. Since the revenues from the natural resources tax are earmarked to
    national, regional and local special budgets, it will be important to
    manage these resources effectively and transparently (for example,
    with an environment fund) so that the additional benefits from
    allocating such moneys are perceived by taxpayers (enterprises and
    local populations). As has occurred in other parts of the world, this
    wise management of the revenues can help to increase acceptance
    of environmental taxes by both business and private citizens.

REFERENCES

Central Statistical Bureau of Latvia (CSBL) (1995a), *Demographic Yearbook of Latvia,* Riga.

Central Statistical Bureau of Latvia (CSBL) (1995b), *A Collection of Statistical Data on Environmental Data, 1994,* Riga.

Central Statistical Bureau of Latvia (CSBL) (1995c), 'Agriculture in Latvia, 1990–1994', in *A Collection of Statistical Data,* Riga.

Central Statistical Bureau of Latvia (CSBL) (1995d), 'Transport and Communications', in *A Collection of Statistical Data,* Riga.

Central Statistical Bureau of Latvia (CSBL) (1995e), *Monthly Bulletin of Latvian Statistics,* No. 9, Riga.

Hahn, Robert W. (1989), 'Economic Prescriptions for Environmental Problems: How the Patient Followed the Doctor's Orders', *Journal of Economic Perspectives,* **3** (2): 95–114.

Ministry of Environmental Protection and Regional Development of Latvia (MEPRDL) (1995), *National Environmental Policy Plan for Latvia,* Accepted by the Cabinet of Ministers of the Republic of Latvia, Riga, 25 April 1995.

State Committee for Statistics of Latvia (SCSL) (1994), *1993 Statistical Yearbook of Latvia,* Riga.

Tietenberg, Tom, H. (1991), 'Economics of Pollution Control: An Overview', in *Environmental and Natural Resource Economics,* 3rd edn, New York: Harper-Collins, pp. 360–89.

Tietenberg, Tom H. (1995), 'Tradable Permits for Pollution Control when Emission Location Matters: What Have We Learned?', *Environmental and Resource Economics,* **5** (2): 95–113.

Weitzman, M. (1974), 'Prices vs. Quantities', *Review of Economic Studies,* **41**: 447–91.

World Bank (1994), 'Infrastructure in Development', *World Development Report 1994,* New York: Oxford University Press.

5. The Lithuanian Pollution Charge System: Evaluation and Prospects for the Future

Daiva Semėnienė, Randall Bluffstone and Linas Čekanavičius

1 OVERVIEW OF ENVIRONMENTAL MANAGEMENT AND THE TRANSITION TO A MARKET ECONOMY

The main goals of the Republic of Lithuania with regard to the environment are first to stem deterioration of the environment and begin to improve its quality, to improve the interaction of human activities and nature and to initiate a transition to sustainable development. Towards these ends, shortly after declaring independence from the Soviet Union in 1990, Lithuania began to focus on several neglected areas. Introduction and improvement of economic instruments was among them (Department of Environmental Protection, 1992).[1]

That economic instruments are an important feature of the environmental policy portfolio is codified in Chapter VI of the Law on Environmental Protection of the Republic of Lithuania, passed in 1992. This law states that the 'ecological and economic interests of the state shall be coordinated by the economic mechanism of environmental protection'. These instruments are defined to include:

- taxes for the use of natural resources;
- charges on pollution;
- economic sanctions and compensation for damages;
- state subsidies;

1 The Department of Environmental Protection was the precursor to the Ministry of Environmental Protection established in 1994.

- price controls;
- credit policies.

The first three instruments are integral parts of the environmental management system. Subsidies for environmental protection in Lithuania are given primarily to the public sector. Only the last two instruments are not in use at all. Other economic instruments used in Lithuania include deposit refund systems for bottles, and fines for noncompliance or wilful misrepresentations.

Although the first three economic instruments are quite important, the Lithuanian system of environmental management is most accurately described as a mixed administrative and economic system. Indeed, it must be emphasized that the main instruments of environmental management in Lithuania are elements of what might be considered a standard system of permits for environmental pollution and the use of natural resources. Components of this system include environmental impact assessment and procedures for granting permission to undertake or ban different kinds of economic activities.

The creation of the system of economic instruments began with the adoption of the Lithuanian Republic Law on Taxes on State-Owned Natural Resources (1991) and shortly thereafter by the Lithuanian Republic Law on Charges on Environmental Pollution (1991). Even before these laws were passed, however, as a first attempt to introduce earmarking in the field of environmental protection, the State Nature Protection Fund was established in 1988.

The goal of pollution charges as described in the law is 'to stimulate pollution abatement and reduce harmful impacts on the environment'. Because of several reasons discussed below, as well as elsewhere (International Development Ireland, 1995a; Semènienè and Kundrotas, 1994; Harrington, 1993), the pollution charge system does not achieve this goal as well as it was designed to do. Nevertheless, pollution charges are in general quite popular, and it is felt that they fulfil an important educational function and also, perhaps, indicate a commitment to the polluter pays principle which can be developed in the future. Moreover, pollution charges raise revenues for the state budget and municipal environmental funds. In 1994, for example, $5.40 million was raised from pollution charges and penalties for above-limit emissions.

The economic context in which any environmental policy reform takes place in Lithuania is at best a complicated one. Manufacturing and mining output declined by 70 per cent during the period 1991–93, and agricultural production, which was approximately 18 per cent of GNP in 1994, fell 26 per cent during the same period (Statistics Department of the

Government of Lithuania, 1995). It is estimated that total output in 1995 was only 30 per cent of output in 1989 (Čekanavičius, 1995). Inflation in 1995 was approximately 35 per cent.

This difficult economic picture, with many enterprises teetering on the verge of bankruptcy, effectively rules out very high-cost solutions to environmental problems. Under such circumstances the cost savings from well-designed economic instruments can potentially be quite important.

The situation is somewhat mixed because any emissions tied to the industrial sector have also declined very significantly. For example, during the period 1989–94, total sulphur dioxide emissions fell by 65 per cent. Particulate and nitrogen oxide emissions fell by similar percentages.[2] Industrial effluents have followed a similar pattern, with chrome emissions falling 75 per cent and copper declining by 67 per cent during 1989–94 (Ministry of Environmental Protection, 1995). These declines in emissions have certainly allowed Lithuania some breathing room, as well as the opportunity to consider relatively less drastic approaches to environmental management.

The rest of this chapter describes the pollution charge/permitting system in Lithuania and attempts to analyse and evaluate that system. Section 2 discusses the permit process and the main environmental problems in Lithuania. Section 3 introduces the charge system and its basic structure, and Section 4 evaluates the charge structure. Conclusions for policy reform are discussed in Section 5.

2   OVERVIEW OF THE POLLUTION PERMITTING SYSTEM IN LITHUANIA

In order to undertake activities in Lithuania which can have harmful effects on the environment, sources must secure a permit from one of the 55 regional departments of the Ministry of Environmental Protection (MEP), or from the MEP if the source is very large. The permit type and procedure is the same whether economic agents extract state-owned mineral resources, such as clay, dolomite, gravel and sand or emit pollution into the environment.

Together with other concerned organizations, regional departments are authorized to review and approve permits for the construction, expansion and operation of any facility. They have the right to take measurements,

2   These developments are at least partly a result of a decrease in energy consumption of 50 per cent which occurred during the same period (Lithuanian Energy Institute, 1994).

collect and analyse samples, and issue enforcement actions. Decisions to issue or not issue permits are based on a review of expected environmental impacts and on the assimilative capacity in a region. Inspectors from these regional departments are responsible for the enforcement of the system and they also work closely with enterprises to set environmental goals, develop control strategies, agree on time frames for abatement, and monitor the progress towards goals. Inspectors can also fine enterprises and in principle can close them down for chronic noncompliance.

Permits for pollution are required for sources which have minimum emission levels defined in the legislation. These polluters may be of any type, including boiler houses, commercial enterprises, industries, municipal agents (such as schools) and so on. For example, any source which uses ten cubic metres of water per day or emits five cubic metres of effluent must have a permit (Bernadišius, 1994).

A central component of the Lithuanian regulatory system is source-level emission standards. For air and water these standards are designed to be ecologically based, and are linked to ambient water and air quality standards. In the case of solid and hazardous waste there are no standards, and permits focus exclusively on amounts and types of wastes which are expected to be generated by sources, and define an agreement regarding the handling and disposal of those wastes.

Ambient standards for air and water quality were determined by the Hygiene Centre of the Republic of Lithuania[3] in 1993, and were taken directly from Soviet-era ambient standards (Zajankauskaitè, 1995). Lithuania still uses these ambient air and water standards, of which there are 800 air and 4,000 water quality standards.[4] The ambient air quality standards specify 30-minute and 24-hour averages for maximum allowable concentrations.

Tables of selected ambient standards from Lithuania and other countries, taken from Table 1 of DHV Consultants (1994a) and Table 6 of DHV Consultants (1994b), are included in the appendix to this chapter. We see from these data that where comparisons are possible it is often the case that Lithuanian standards are stricter than those in force in other countries or suggested by the World Health Organization.

As of 1995 this system of environmental standards was under revision and it was expected that new standards would be in effect during 1996 or 1997. It is, for example, foreseen that more stringent standards will be

---

3 The Hygiene Centre is the public health arm of the Ministry of Health and is responsible for setting sanitary and other standards in order to adequately protect human health.

4 In cases where ambient standards have not been determined, enterprises must apply to various scientific organizations to establish ambient standards before emitting pollutants into the air or water.

assigned to 'sensitive' waters, those which serve special functions, unpolluted 'natural' waters and waters in sensitive areas, national parks or nature reserves. Less strict standards will be assigned to heavily polluted waters and to areas with a high concentration of polluters.

The standard applied to a particular new or existing source depends directly on the ambient standard in force, as well as on the assimilative capacity of the environment and the number and composition of pollution sources in the area for which the permit is requested. These factors are summarized in formulas codified in regulations which enterprises must use to calculate the maximum amount they can emit without causing ambient standards to be violated.

In the case of water pollution, ambient standards are set to allow surface waters to be suitable to maintain river ecosystems. They are also set to comply with Baltic Marine Environment Commission (HELCOM) recommendations for most pollutants.[5] The relationship between the maximum ambient concentration for a particular water pollutant and the permitted effluent level for a source is made using four formulas given to the polluter by the MEP. The basic idea is that, given the characteristics of a waterway, a maximum contribution to ambient concentrations is assigned to the marginal source such that water quality remains at or above a cutoff level. These concentration-based measures are then converted to maximum permitted emissions per unit of time. Important factors determining permitted effluent levels include:

- existing concentration of pollutants in a river or stream at a measurement point;
- turbulence of the waterway;
- maximum permissible concentration at a measurement point.

Air polluters requesting a permit are also assigned standards which do not cause ambient concentrations to exceed the maximum permitted concentration value in that region. This maximum ambient concentration is defined as 'a . . . concentration of [the] chemical substance in the air in human settlements . . . exhibiting no reaction to it after 20–30 minutes [of exposure]' (Department of Environmental Protection, 1992, p. 40). To determine whether ambient standards will be exceeded by permitting a source for a given discharge, a computerized air dispersion model is used. It has been recommended that these models be updated

5   Recommendations are standards for BOD_5 and suspended solids. In practice it was acknowledged early on that HELCOM recommendations were unlikely to be met before 1995. Higher ambient standards were therefore set to define acceptably treated water (Department of Environmental Protection, 1992).

using Western models, but because of difficulties in testing the model the extent of any inaccuracies is not known (Teriošina, 1995; USEPA, 1995).

Polluting enterprises have substantial input into the permit process and work very closely with regulators every step of the way. For example, the first step of the process for air polluters is to draw up inventories of pollution sources themselves, often relying on consulting companies or universities to conduct this inventory. This step provides the background for all further regulatory work related to that enterprise. Sources themselves then prepare draft permits covering a period not longer than five years. In these draft permits they propose their own emission standards and document that ambient standards will not be violated.

In the case of water pollution, sources use the four formulas mentioned above to set their standards. For air pollution, enterprises request data on the existing background concentration from the MEP and then make a permit proposal which covers all their sources. These permit proposals include proposed aggregate emission levels, which will keep overall air quality within the required limits. Standards are also proposed for each source.

It is notable that these draft air pollution permits often include multiple sources and enterprises can in principle operate like an air pollution bubble. For example, an enterprise may propose an increase in emissions from one source as long as the dispersion model shows that another source has reduced emissions sufficiently to contribute to meeting ambient air quality goals. Trades across enterprises in principle are also possible, but as of 1995 it had not been documented that this occurred (Teriošina, 1995).

The emission standards applied to each source are called the maximum allowable pollution (MAP) level and in principle apply to polluters of any significant size. In practice, however, many enterprises at least claim not to be able to meet MAP standards and are able to convince MEP regional departments that they deserve less stringent standards associated with temporary allowable pollution (TAP) permits. Emission levels specified in TAP permits depend exclusively on negotiations between MEP inspectors and pollution sources regarding enterprises' financial and technical capabilities to meet MAP standards, and on the existing economic and technical circumstances. TAP permits have a maximum duration of five years and specify steps polluters must take to allow them to meet MAP standards.

How frequently are TAP permits given? In fact, TAP permits are the typical case for water polluters. In a random sample of 366 sources meeting the criteria requiring them to calculate their MAP levels in 1994, for those

Table 5.1   *Percentage of total permit holders operating under temporary allowable pollution permits in 1994*

	Air Pollutants				Water Pollutants			
	SO_2	NO_x	Pb	M_nO_x	BOD_5	Solids	Oil	N
Percentage	36.7%	37.6%	37%	11.7%	74.1%	70.0%	73.4%	76.4%
Total permit holders	193	231	27	102	147	160	113	72

Source: Random Sample of 366 Polluters; Harvard Institute for International Development and Ministry of Environmental Protection (1995).

emitting the four most common water pollutants at most 23.6 per cent were operating under MAP standards.[6] These data are presented in Table 5.1.

In the case of air pollution, a much larger percentage of sources operate under MAP permits. The most likely reason for this difference is that air quality is less of a problem than water quality, and for relatively few pollutants are ambient standards exceeded in urban areas. The Lithuanian National Environmental Strategy Background Report, for example, concluded that 'The results of ambient air quality measurements do not indicate that substantial and expensive air pollution control equipment for stationary emission sources is justified' (International Development Ireland, 1995a, pp. 3–1). Compared with other Central and Eastern European countries, Lithuania's air pollution problem appears to be quite minor.[7]

Water quality, however, is considered to be a major problem. In approximately 80–90 per cent of Lithuanian rivers, BOD_5 levels are above ambient standards and 20 per cent are more than twice the standard. As shown in Appendix 5A.2, this seemingly poor performance would still be well within European Union water quality standards.

6   There are approximately 4,000 polluters in Lithunia who are controlled by the MEP. Water polluters include both municipal wastewater treatment plants and enterprises discharging directly into waterways.
7   Although automobile ownership in Lithuania is relatively low at 155 passenger cars per 1,000 population at the end of 1993 (Statistics Department of the Government of Lithuania, 1995), approximately 60 per cent of all air emissions are from mobile sources. In the US, ownership of passenger cars was more than three times as high in 1990 (Anderson, 1990).

A major problem is a lack of wastewater treatment facilities. In 1993, for example, of wastewater requiring treatment, 23.0 per cent of discharges received no treatment at all. Only 26 per cent of discharges were of a sufficient quality to meet ambient standards (Ministry of Environmental Protection, 1995; International Development Ireland, 1995a). Water quality is expected to improve dramatically by the end of the century, however, when wastewater treatment facilities will be completed in the five major Lithuanian cities. Until that time the water quality management system is somewhat on hold, because these major discharges put so much pressure on surface water quality that it is difficult or impossible for other sources to meet their MAP standards.

Perhaps the second most important reason for the difference in incidence of TAP permits is that stationary source air emissions are closely linked with industrial activity, which has declined dramatically since 1990, while water quality is most affected by amounts of sewage and how it is treated. Over time, achievement of MAP standards has become easier for air polluters than for major dischargers of water pollution such as municipal wastewater treatment plants.

Monitoring is the responsibility of each enterprise, but continuous emission monitoring equipment is not required. Enterprises are responsible for keeping records of their emissions and for sending results to MEP regional departments on a quarterly basis. It is on the basis of these reports that charges are assessed. Fines for misreporting emissions are ten times what would have otherwise been paid.

This self-monitoring system is enforced by spot checks by inspectors. In most cases these visits are unannounced. Each inspector covers several plants and measurement equipment in the MEP is considered to be outdated (USEPA, 1995). Disagreements over MEP calculations are said to be quite common.

3   THE SCOPE AND STRUCTURE OF CHARGES IN LITHUANIA

The Law on Taxes of Environmental Pollution (April 1991) states that the goal of imposing pollution taxes is to 'serve as an economic element of environmental protection which stimulate[s] pollution abatement and reduce[s] the harmful impact on the environment'. The goal of the legislation is therefore to influence the level of emissions by polluters (Department of Environmental Protection, 1992). A more precise goal has not been established. For example, the system of water pollution charges could potentially be calibrated to assist the country to meet HELCOM water quality recommendations (Bluffstone, 1995).

To serve this purpose, the Lithuanian government developed a complex system of pollution charges which in principle covers all pollutants. Pollution charges are levied for air pollution (more than 100 different pollutants) and water pollution (51 pollutants). A partial list of charge rates are presented in Appendix 5A.3. As of 1995, charges were still assessed individually on that number of pollutants, although plans were under way to simplify the system by reducing the number of categories of pollutants subject to taxation.

Total earnings from the system of pollution charges and penalties were 21.6 million litas (Lt.) (US$ 5.4 million) in 1994. With the exception of pollution charge collections from a few large plants, which may have different distributions, 70 per cent of collected charges are allocated to municipal government environmental funds (total of 55). Average annual contributions to these funds are about $70,000, making accumulation of sufficient funds to make investments difficult (CowiConsult, 1995).[8] The remaining 30 per cent goes to the state budget. Penalties for exceeding standards are accumulated in the State Nature Protection Fund. These funds are then used for environmental investments and incentives for inspectors.

The developers of the Lithuanian charge system attempted to create a charge rate structure which mimicked an increasing marginal damage function. It is therefore very difficult to talk about "the" charge rate because, particularly for sources emitting under an MAP permit, the unit charge is always increasing in polluters' emissions levels. Perhaps the most important implication of this setup is that in most cases the *rate* an enterprise pays for emissions of a given pollutant will be different from that paid by any another enterprise.

Within this structure, there are three levels of charges. Base rates were established in the original legislation and are indexed quarterly. There are also penalty rates when emissions exceed standards and preferential rates when emissions are below MAP standards. Both of these rates are linear functions of base rates and emission levels, with total penalties pro-rated depending upon the amount of time standards are actually exceeded.

Charges are estimated at the beginning of each year, assuming that enterprises exactly achieve their standards, and are paid in advance on a quarterly basis. A settling of accounts occurs in the fourth quarter so that by the end of the year owed charges and penalties are equivalent to those which have been paid.

8   Serious questions have also been raised about the degree to which these funds are actually benefiting the environment.

An enterprise's charge rate for *all units* it emits, and therefore its total charge payment, is determined by comparing its actual emissions with its permitted standard. For example, if a source operates under an MAP standard and emits a pollutant at a level less than 50 per cent of its standard it pays *no* pollution charges. If this same source cannot reach that level of reduction, but still emits below its standard, the firm still receives a reward in the form of a preferential rate valid for *all* units it emits. If we define T_i as the total tax paid by the *i*-th source, T_0 as the base rate, F_i as the total emissions of source *i* and N_i as the standard, MAP permit holders above 50 per cent, but below 100 per cent, of permitted standards pay total charges of:

$$T_i = F_i * T_0 * (1 - 2 * (N_i - F_i) / N_i).$$

Sources with TAP permits who emit below their standard, as well as both TAP and MAP permit holders exactly *at* their standards, pay the base rate specified in Appendix 5A.3. These sources therefore pay total charges equal to:[9]

$$T_i = T_0 * F_i.$$

For sources who are above their standards, a penalty rate applies. It is particularly important to note that this higher rate is levied on *all* units emitted by sources and not just on those above the permitted level. This rate structure therefore has the potential to create very high marginal penalty rates and therefore substantial incentives for sources to keep their emissions at or below their MAP or TAP permitted levels.

There is some differentiation in penalty rates depending on the type of enterprise being penalized. Water polluters and power plants pay total penalties of:

$$T_i = T_0 * F_i * (1 + F_i / N_i).$$

If an enterprise is defined as an industrial air polluter, however, the penalty rate is at least three times higher. This group pays penalties of:

$$T_i = T_0 * F_i * (1 + 4F_i / N_i).$$

---

9   Air pollution charges are also paid for fleets of vehicles using the base rate applied to the actual emissions of vehicles. A similar system is used to charge for stormwater runoff.

Total charges resulting from this system are shown in Figure 5.1 for a medium-sized, industrial air polluter emitting sulphur dioxide under an MAP permit. Under the Lithuanian system, total charges are an increasing function of emissions and they rise at an increasing rate. Total charges increase particularly sharply after the source crosses its emission standard. For example, emitting 290 tons instead of its standard of 216.143 tons increases the firm's pollution charge burden by more than fivefold.

Charge waivers are also available for those polluters who operate under MAP permits and who also implement pollution-abatement measures that are designed to reduce pollution by 25 per cent. The limit on the waiver is the lesser amount of the total investment cost and the amount of pollution charges which would have been paid over three years without the waiver.

In practice, this waiver is used at most by 0–2 per cent of emitters of major pollutants (Harvard Institute for International Development and MEP, 1995). As will be discussed below, charge payments are very low for most enterprises, making waivers of pollution charges a relatively ineffective instrument. More than 70 per cent of water polluters operate under TAP permits and are specifically excluded from this programme.

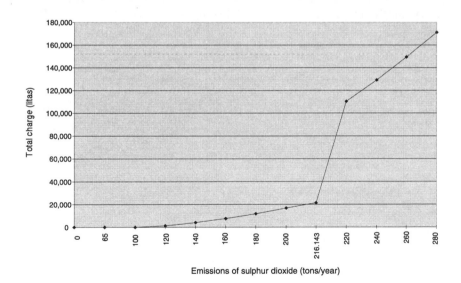

Figure 5.1   Total charge payment schedule for a sample industrial sulphur-dioxide polluter in Lithuania in 1994

4 ANALYSIS OF THE CHARGE SYSTEM

In setting base charge rates, significant attention was given to the environmental damage which different pollutants were expected to cause. Charges are therefore differentiated according to environmental damages, but it is unlikely that the *levels* of these charges in any way reflect the damage caused by these pollutants. Indeed, little economic analysis was conducted in setting charge rates (Semėnienė and Kundrotas, 1994).[10] In particular, there is little relation between charge rates and costs of abatement.

As shown in Table 5.2, total annual charge payments in 1994 were very modest to say the least. For polluters paying the median level of charges for emissions of ten major pollutants, total charges ranged from about Lt. 14 ($3.50) to Lt. 159 ($39.75) per pollutant. Indeed, if a hypothetical polluter emitted all ten pollutants and paid the median level of charges for all ten, the total bill would be only Lt. 628.72 ($157.18) per year. Though it should be emphasized that even low levels of charges can have *some* incentive effects if there is a possibility for sources to receive credit for pollution abatement, it is hard to imagine anything but virtually no-cost abatement measures being stimulated by these levels of charges.

As shown in the table of base charge rates in Appendix 5A.3, in 1995 the rules of indexation were altered to fully index charge rates for the effects of inflation. This revision was done retroactively, and from the second to third quarters charge rates were doubled for most pollutants. This change somewhat increased incentives for abatement, but charge payments still remained extremely low.

Penalties are also rarely applied. As shown in Table 5.2, for the ten pollutants presented, out of a sample of 366 polluters only fifty penalties were levied in 1994,[11] which is extremely low considering that penalties can be applied and pro-rated for violations of firm-level standards occurring during any portion of a year. When penalties are applied, however, they can be quite high compared with charge levels.

Because penalties are only applied when sources emit over their standards, there is a strong link between the observed tendency to grant

10 It should be noted, however, that a perhaps imperfect substitute called 'Temporary Methods to Calculate the Economical Efficiency of Implementation of Environmental Measures and to Estimate the Damage Suffered by the National Economy Due to Environmental Pollution', by the Academy of Sciences of the USSR (1983) was used to set base pollution rates.
11 It should be recalled that each polluter typically emits several pollutants. Total possible penalties were 1,262, implying a penalty rate of about 4.0 per cent.

Table 5.2   Charges and penalties paid by air and water polluters in 1994

Air pollutants	NO_x	SO_2	Pb	Organic/ inorganic dust	M_nO_x
Base rate (Lt./ton)	98.94	52.83	53,793.6	47.07	16,978.6
Median total charges paid (Lt.)	159.06	117.38	47.94	36.24	15.13
Median total penalties paid (Lt.)	148.63	4,015	None	102.87	131.39
Number of penalties	5	3	0	7	3

Water pollutants	P	N	Oil	BOD_5	Suspended solids
Base rate (Lt./ton)	378.96	111.43	4,457.18	111.43	22.09
Median total charges paid (Lt.)	37.58	47.65	103.28	49.65	14.82
Median total penalties paid (Lt.)	696.24	93.07	207.66	806.06	65.45
Number of penalties	2	3	3	10	14

Note: $1.00 = Lt. 4.00

Source: Random Sample of 366 Polluters; Harvard Institute for International Development and Ministry of Environmental Protection (1995).

less stringent TAP permits and the prevalence of penalties. If more sources were held to MAP standards the incidence of penalties should be greater. Indeed, business leaders report that when it is clear that polluters cannot meet their MAP standards, rather than holding them to those standards and subjecting them to penalties, a TAP permit is issued for the actual emissions (Ecological Engineering Association of Lithuania, 1995).

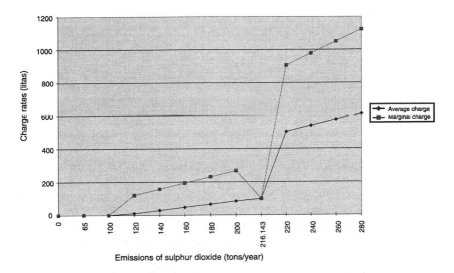

Figure 5.2   Average and marginal charge rates for a sample industrial polluter of sulphur dioxide in 1994

It is also notable that relatively few companies pay the vast majority of pollution charges in Lithuania. For example, the largest 2.3 per cent of emitters of sulphur dioxide pay 66 per cent of the charges on that pollutant, and the top 2.8 per cent of dust polluters pay 71 per cent of total charges. The story is similar for other air emissions and for water pollutants (Harvard Institute for International Development, 1995).[12]

An examination of the rate structure chosen in Lithuania reveals a variety of features. Figure 5.2 presents the average and marginal charges facing an enterprise. It is particularly notable that marginal charges are generally rising, except for a sharp decline exactly at the standard.

As noted above, charge payments by pollution sources are low and are therefore unlikely to encourage much in the way of environmental investments and probably do not even stimulate low-cost changes in production processes. Consequently, the effect of the pollution charge component of the system on behaviour is likely to be relatively small; thus the extent that charges are well or poorly constructed probably has only a minimal impact. Because this conclusion could easily change if the base charge

12   For example, 2.2 per cent of BOD_5 polluters pay 62 per cent of all charges and 4.4 per cent of phosphorus and nitrogen polluters pay, respectively, 70 per cent and 50 per cent of all charges.

rates are increased, it is perhaps nevertheless useful to evaluate the efficiency properties of the Lithuanian charge structure, as well as the overall mixed system.

From profit maximization we know that enterprises will voluntarily choose emissions such that marginal abatement costs equal marginal charges imposed by regulators. In Figure 5.3, hypothetical marginal abatement costs are plotted along with the marginal charge from Figure 5.2.

For simplicity it is supposed that all firms have a permitted MAP standard for sulphur dioxide of 216.4 tons per year. This perhaps over-simplifies, because in reality standards differ across enterprises and the marginal charge schedule in Figure 5.2 would therefore shift to the right or left depending on whether standards are above or below 216.4 tons per year. As expected, we see from Figure 5.3 that different abatement costs imply that equilibrium emission levels differ across firms, and may be above, below or equal to standards depending on costs.

Marginal abatement costs will also be different across pollution sources. This is potentially important, because it is only when marginal costs are equal across firms that an economic instrument, such as a pollution charge, minimizes the aggregate cost of pollution abatement (Baumol and Oates, 1988). Indeed, in the neighbourhood of the standard, marginal charges are greater by a factor of three below the standard and

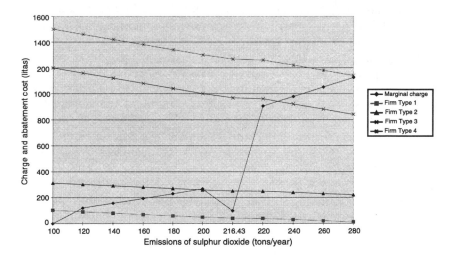

Figure 5.3   Marginal sulphur-dioxide charge and possible marginal abatement cost schedules of enterprises with identical standards

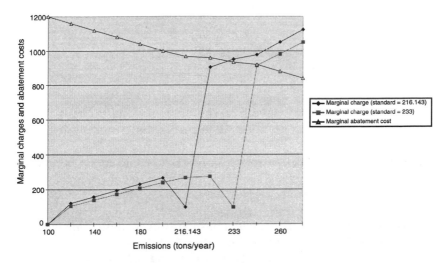

Figure 5.4   Marginal charge schedules for two sulphur-dioxide polluters with different standards but the same marginal abatement cost schedules in 1994

by a factor of nine as firms incur penalties. With marginal penalties continuously increasing above the standard, this divergence of marginal abatement costs merely increases.

Differences in penalty rates for different types of polluters (that is, industrial air polluters and all others) merely exacerbate this inefficiency. At least in a world of perfect information,[13] therefore, this charge structure creates a relatively inefficient instrument.

The progressive nature of the marginal charge function is certainly an important source of economic inefficiency, but perhaps the most important factor is the system of source-specific standards. This problem is illustrated in Figure 5.4 for two enterprises with identical marginal abatement costs. Even if enterprises are identical in terms of their abatement options, because charges are assessed with reference to source-specific standards, if their standards differ they will choose different equilibrium emission levels and have different marginal abatement costs.

The combination of charges and standards also implies that for virtually all situations only firm-specific analysis is possible. In particular,

13   Meaning that abatement costs and damages are known and monitoring by the MEP is perfect.

aggregation of firm-specific marginal cost curves into industry or 'polluter' marginal cost curves becomes very difficult, and predicting the effects of a charge schedule is also complicated.

5   CONCLUSIONS AND DIRECTIONS FOR POLICY REFORM UNDER UNCERTAINTY

As noted in Section 1, given the tenuous economic situation during the transition to a developed market economy, high-cost environmental policies are non-starters. Well-designed economic instruments therefore are particularly of interest, and since pollution charges are the main economic instrument in Lithuania, it is perhaps worth considering changes which should be made in the system.

In a mixed system of firm-level standards and pollution charges, only one of these instruments will be the primary determinant of enterprise behaviour; if charges are very low, standards will be binding, and if charges are high emissions will be below standards (Bluffstone and Farrow, 1995). To the extent that enterprises are abating pollution at all in Lithuania, they are likely to focus on achieving – or appearing to achieve – their standards because charge payments are so low. For those who are unable to achieve their standards, the penalty portion of the charge structure could potentially function as an 'economic' instrument in the sense that it may influence the behaviour of polluters. Under such circumstances, the problems with cost effectiveness are the typical ones associated with standards. Perhaps the most general statement which can be made about the charge system in Lithuania is therefore that it does little damage.

Recognizing, as was done almost 25 years ago by Baumol and Oates (1971), that we at best live in a world where cost effectiveness, rather than true optimality in the Pigouvian sense, is the goal of pollution charges, what can be said about the cost effectiveness of the overall charge/standard system? Is the system a particularly bad one? The answer is probably no, for two reasons: (a) a substantial portion of the permitting system is tailored to individual enterprise characteristics; and (b) the real situation is one in which there is substantial uncertainty about marginal abatement costs and enforcement effectiveness.

With regard to the first point, TAP permits are given on the basis of perceived marginal abatement costs and the financial conditions of each enterprise. Although there is likely to be substantial slippage in the Lithuanian permitting system because of its informality, a conscious effort appears to be made to take the costs of abatement into account when standards are set. As discussed by Oates et al. (1989) in their study of particulate emissions in Baltimore, Maryland, such an approach can mimic pretty well the important aspect of flexibility which is valued in economic instruments. Transferring these empirical results from the United States to Lithuania should, of course, be viewed with caution, but it does raise a question regarding the degree of improvement in cost effectiveness which can be expected from an increased reliance on pollution charges.

Perhaps of relevance, also, is the well-developed literature on optimal instrument choice. Beginning with Weitzman (1974) and Adar and Griffin (1976) it was recognized that when damages rise very quickly compared with abatement costs, emphasis should be on controlling *quantities* of pollution rather than *prices* of pollution. Allowing standards to be transferable within an appropriate region increases the cost effectiveness of such a standards-based approach. For many pollutants controlled by the MEP, this condition is likely to hold and a primary emphasis on standards is probably justified.

This brings us to the second point, uncertainty. To set pollution charges properly one must know something about marginal abatement costs, but the MEP certainly lacks accurate information on abatement costs and this problem is not peculiar to Lithuania. The applied literature focusing on calculation of marginal abatement costs, particularly when several industries and pollutants are involved, is extremely small and there are few 'off-the-shelf' studies to assist the MEP in estimating compliance costs.[14]

A second source of uncertainty regards the distinction between actual and reported emissions. As discussed above, monitoring is based on self-reporting in Lithuania, with fines levied for understating emissions. Inspectors make spot checks on average one to four times per year. How effective is the Lithuanian charge/standard system in dealing with uncertainties in emissions monitoring?

An important way to mitigate this type of uncertainty is to create incentives for developing identifiable environmental 'projects'. The Lithuanian system, in principle, offers such incentives by providing environmental investment subsidies through rebates of pollution charges when emissions decline by at least 25 per cent from the previous year's level. An environmental fund will also be created in 1996, which will provide access to credit for a variety of investments having environ-

14 An exception is Hartman et al. (1994) which contains a very detailed analysis of US data.

mental benefits. Both these instruments create projects for which emissions can be relatively accurately estimated from engineering information. There are also tenfold fines for misrepresenting emissions, which is an important component of an effective system (Harford, 1978).

Can the system be improved? The answer is certainly yes. Perhaps most fundamentally, the charge system is much too complicated and this complexity probably contributes little to achieving the environmental goals of the Republic. However, to choose the correct way to simplify the system, a more precise vision of the role of charges in the overall environmental policy is required. An important related question arises whether it is desirable to raise charges so that they affect the behaviour of firms, rather than merely provide a complicated, but well-defined system for penalizing polluters who are above their standards. This question alone generates a host of other issues regarding the uses to which those revenues should be put and whether any offsetting compensation, either in the form of relief from other taxes or redistribution of charges, should be given to polluters.

If charges are to be a primary instrument for influencing emissions of certain pollutants, it would also be useful to articulate more precisely the environmental goals the mixed system in Lithuania is supposed to hit. With regard to water pollution, for example, if the goal is to support achievement of HELCOM recommendations, this objective could be made more explicit. More precisely defining the environmental goal would then provide aggregate emissions targets which the system could be calibrated to achieve.

The system of ambient standards is in need of revision, particularly where standards exceed those in Western Europe, because overly strict ambient requirements put excessive compliance burdens on enterprises. The MEP began a revision of these standards in 1995 and a new system will be proposed in 1996 or 1997.

The structure of the charge system should probably be revised by creating a standard two-tiered charge structure in which rates do not depend on emissions. This structure would eliminate most of the inefficiencies associated with the current charge structure. Under such a revised structure there would be two rates, one which is a base rate and one which is a penalty rate paid only on the emissions above firms' standards. An estimated uniform charge rate which is expected to achieve the stated environmental goals would presumably fall somewhere in between these two rates. Such a structure does a better job than a uniform charge when abatement costs are uncertain. Costs are indeed minimized if standards are transferable (Roberts and Spence, 1976).

Charge rates will probably need to be increased whether charges are calibrated to enterprises' abatement costs (therefore becoming a binding policy instrument) or whether they will mainly support standards by generating revenues and fulfilling a penalty function. Increased rates will provide incentives for those who can easily meet their standards to do even better. For those above standards, more meaningful charge rates will provide a more effective economic instrument to encourage firms to reduce emissions to levels at or below their standards. Raising charge payments would also provide an important incentive for firms to take advantage of the pollution charge waiver option. This is potentially important, because it means that MEP regional departments will have approved a well-defined project and will therefore have good estimates of the emission reductions to be expected; use of this subsidy option therefore economizes on monitoring costs.

If an increased charge revenue stream is contemplated, it will be necessary to ensure that it is well spent, and probably that means channelling some of those resources back to enterprises. Substitution for other taxes is a possible, though unlikely, option within the overall search for a balanced government budget in Lithuania. Perhaps a more feasible way to funnel pollution charges back to firms is to use the environmental fund which will be operational in 1996. This vehicle might also be combined with expanded use of the pollution charge waiver.

Another way the financial burden on enterprises can be reduced is by rebating the unused portion of estimated charges after each quarter rather than at the end of the year. In the high-inflation environment found in Lithuania this is important, because such advance payments effectively amount to a large hidden tax. Moreover, this system of advanced payments and rebates reduces abatement incentives, because credit for any reductions in emissions are received only at the end of the year when they may be worth 30–40 per cent less than at the beginning of the year.

Given the importance of standards in the Lithuanian system, performance could almost certainly be improved by allowing trading in standards across pollution sources and across firms. The MEP has a conceptual structure on which it can build in this area, because the system of air emission regulations allows for bubbling within a firm and even across firms. In the jargon of the Lithuanian system, this is simply allowing the standards on some sources to be relaxed as long as they can show that stricter standards on other sources will improve air quality by a greater amount than without the 'trade'.

The MEP should consider encouraging this aspect of the system by making it more explicit and by promoting it as a way to reduce costs. The burden of calculating standards based on ambient environmental conditions is already placed on enterprises. It would therefore be extremely easy to let firms propose slightly more complicated regulatory programmes which involve bubbling of sources within enterprises, or netting of emissions with other sources. Such a system could also potentially be extended to water pollution, either by promoting the notion with municipal water companies looking for ways to reduce compliance costs of subscribers, or by allowing trades among enterprises who are close to each other and who discharge directly into waterways (Harrington, 1993).

REFERENCES

Adar, Zvi and James Griffin (1976), 'Uncertainty and the Choice of Pollution Control Instruments', *Journal of Environmental Economics and Management*, **3** (3) October: 178–88.

Anderson, Robert (1990), 'Reducing Emissions from Older Vehicles', *Research Study No. 53,* American Petroleum Institute.

Baumol, William and Wallace Oates (1971), 'The Use of Standards and Prices for Protection of the Environment', *Swedish Journal of Economics,* **73** March: 42–54.

Baumol, William and Wallace Oates (1988), *The Theory of Environmental Policy,* 2nd edn, Cambridge: Cambridge University Press.

Bernadišius, Vytautas (1994), 'Water Resources and Protection Policy in the Republic of Lithuania', Unpublished Working Paper of the Ministry of Environmental Protection of the Republic of Lithuania.

Bluffstone, Randall (1995), 'Achieving Environmental and Fiscal Goals in Lithuania Using Pollution Charges', Working Paper, Harvard Institute for International Development, Harvard University.

Bluffstone, Randall and Scott Farrow (1995), 'Adapting Economic Instruments Within Existing Administrative Systems: The Case of Air Pollution in Northern Bohemia', Mimeo.

Čekanavičius, Linas (1995), 'Proposal for Environmental Policy Reform in Lithuania: A Transferable Universal Emission Coupons Scheme', Mimeo.

CowiConsult (1995), 'Establishment of a New Environmental Revolving Fund', Prepared for the Ministry of Environmental Protection and the European Commission.

Department of Environmental Protection of the Republic of Lithuania (1991), 'The Lithuanian Republic Law on Taxes on Environmental Pollution', Passed 2 April 1991 (Translated and Included in Information Bulletin No. 1 (1992) of the Department of Environmental Protection), pp. 24–39.

Department of Environmental Protection of the Republic of Lithuania (1992), 'Law on Environmental Protection of the Republic of Lithuania', Passed 21 January 1992 (Translated and Included in Information Bulletin No. 1 (1992) of the Department of Environmental Protection), pp. 5–19.

DHV Consultants (1994a), 'Basic Document on Proposed Standards for Ambient Air and Industrial Emissions', Report to EU–Phare and the Environmental Protection Department of the Republic of Lithuania.

DHV Consultants (1994b), 'Water Quality Standards in Lithuania', Report to EU–Phare and the Environmental Protection Department of the Republic of Lithuania.

Ecological Engineering Association of Lithuania (1995), Private Communication with Rimantas Budrys, President, January.

Harford, Jon (1978), 'Firm Behavior Under Imperfectly Enforceable Pollution Standards and Taxes', *Journal of Environmental Economics and Management*, **5** (1) March: 26–43.

Harrington, Winston (1993), 'Air and Water Quality Permitting in Lithuania', Working Paper, Harvard Institute for International Development, Harvard University.

Hartman, Raymond, Manjula Singh and David Wheeler (1994), 'The Cost of Air Pollution Abatement', *World Bank Policy Research Working Paper 1398*, Washington, DC: World Bank.

Harvard Institute for International Development and Ministry of Environmental Protection of the Republic of Lithuania (1995), 'Sample of Lithuanian Polluters', Mimeo.

International Development Ireland, Ltd. and Environmental Resource Management, Ltd. (1995a), 'National Environmental Strategy of the Republic of Lithuania: Background to the Strategy', March.

International Development Ireland, Ltd. and Environmental Resource Management, Ltd. (1995b), 'National Environmental Strategy of the Republic of Lithuania: Final Report', September.

Lithuanian Energy Institute (1994), *Energy In Lithuania*, Kaunas, Lithuania.

Ministry of Environmental Protection (1995), *Lithuania's Environment: Status, Processes, Trends*, Vilnius, Lithuania.

Oates, Wallace, Paul Portney and Albert McGartland (1989), 'The Net Benefits of Incentive-Based Regulation: The Case of Environmental Standard-Setting in the Real World', *Discussion Paper CRM 89–03*, March Resources for the Future.

Roberts, M.J. and Spence (1976), 'Effluent Charges and License Under Uncertainty', *Journal of Public Economics*, **5** April/May: 193–208.

Semėnienė, Daiva and Arunas Kundrotas (1994), 'From Theory to Practice: Environmental Taxes in a Time of Economic Uncertainty', Paper Presented at the 50th Congress of International Institute of Public Finance, Cambridge, MA.

Statistics Department of the Government of Lithuania (1995), *Lithuania: Selected Data 1993–1994*, Vilnius, Lithuania.

Teriošina, Marija (1995), Air Pollution Division of the Ministry of Environmental Protection, Private Communication, August.

US Environmental Protection Agency (USEPA) (1995), 'Baltic Republics' Environmental Monitoring Review Project: Findings and Recommendations, Lithuania', EP 905–R–95–005.

Weitzman, Martin (1974), 'Prices vs. Quantities', *Review of Economic Studies*, **41**: 447–91.

Zajankauskaitė, Juratė (1995), 'Use of Natural Resources: An Overview of the Process of Issuing and Implementing the Pollution and Natural Resource Permit System in Lithuania', Working Paper, Harvard Institute for International Development, Harvard University.

APPENDIX 5A

Table 5A.1 Ambient air quality standards for selected air pollutants in Lithuania and other countries

Substance	Average	Ambient standard in micrograms/m³				
		Lithuania	EU	WHO	Netherlands	Germany
Sulphur dioxide (SO_2)	30 minutes	500	–	–	830	–
	24 hours	50	100–150	125	500	–
	1 year	–	80–120	50	75	140
Nitrogen oxides (NO_x/NO_2)	30 minutes	400/85	–	–	–	–
	24 hours	60/40	–	400	–	–
	1 year	–	200	150	135	80
Carbon monoxide (CO)	30 minutes	5,000	–	60,000	40,000	–
	24 hours	3,000	–	10,000	6,000	10,000
	1 year	–	–	–	–	–
Volatile organic compounds (VOC)	30 minutes	–	–	–	–	–
	24 hours	–	–	–	–	–
	1 year	–	–	–	–	–

Total suspended particles (TSP)	30 minutes	500	–	–	–	–
	24 hours	150	–	–	140	–
	1 year	–	–	120	40	150
Black smoke	30 minutes	150	–	–	–	–
	24 hours	50	100–150	125	150	–
	1 year	–	80	50	–	–
Fluoride (F) and its compounds	30 minutes	20	–	–	–	–
	24 hours	5	–	–	–	–
	1 year	–	–	–	2.8	1
Chlorine and/or hydrogen chloride	30 minutes	100/200	–	–	–	–
	24 hours	30/200	–	–	–	–
	1 year	–	–	–	–	100
Ozone	30 minutes	160	–	200	240	–
	24 hours	30	–	–	160	–
	1 year	–	–	–	–	–

125

Source: DHV Consultants (1994a, Table 1).

Table 5A.2.   *Ambient Water Quality Standards for Selected Water Pollutants in Lithuania and Other Countries*

Water Quality Standards	Values in Micrograms/liter or as indicated				
	Lithuania	US	Netherlands	EU	Canada
Oxygen (mg/l)	≥ 6		≥ 5	100% ≥ 5 50% ≥ 8	5–9.5
pH	6.5–8.5		6.5–9	6–9	6.5–9
BOD$_5$ (mg O$_2$/l)	2		–	6	
Total Phosphorus (mg P/l)	0.2*		0.15		
Total Nitrogen (mg N/l)	2*		2.2		
Ammonia (NH$_3$) (mg N/l)	0.04		0.02	0.004	
Chlorine (residual)	0	–	–	5	2
Chorides (Cl$^-$) (mg Cl/l)	300		200		
Fluorides (F$^-$) (mg F/l)	0.5*		1.5		0.3
Sulfates (SO$_4{}^{2-}$) (mg SO$_4$/l)	100		100		
Cadmium (Cd)	5	10	0.05	5	0.2–1.8
Mercury (Hg)	0.01	0.14	0.02		0.1
Nickel (Nl)	10	13	9		25–150
Lead (Pb)	100	50	4		1–7
Iron (Fe)	100	300			
Phenols	1	350	2		1
DDT-total	0	0.024	**	25	0.001

Notes: *) Temporary proposal
         **) A limit has only been set for sediment, being the more appropriate place for measurement

Source: DHV Consultants (1994b)

Table 5A.3.   *Selected Basic Charge Rates for Water and Air Pollution in Lithuania in 1995*

Selected Water Pollution Basic Charge Rates

Pollutant	Charge Rates Second Quarter 1995 (Lt./Ton)	Charge Rates Third Quarter1 995 (Lt./Ton)
BOD$_5$	148	301
Suspended Solids	29	60
Oil Products	5908	12,055
Phenols	295,382	602,736
Phosphorus	502	1025
Nitrogen	148	301
Mercury	2,953,820	6,027,360
Cadmium	59,077	120,547
Chrome	29,538	60,274
Lead	22,480	45,870

Selected Air Pollution Basic Charge Rates

Nitrogen Oxides	131	268
Sulfur Dioxide	70	143
Carbon Monoxide	3	7
3,4 Benzo (a) pirene	4,010,580	8,184,000
Dust (Various Kinds)	64-223	127-455

6. Pollution Charges in Russia: The Experience of 1990–1995

Michael Kozeltsev and Anil Markandya

1 INTRODUCTION

The government of the former Soviet Union began paying more attention to environmental issues in the early 1970s. At that stage, and until its collapse in the 1990s, a system of centralized management was the chief mechanism of environmental regulation, using mostly command-and-control methods. The highest point in this system was the introduction of nature protection plans in 1982 as an integral part of the overall planning process.

In the 1980s, the government experimented with economic incentives in the field of environmental management, starting with payments for the use of natural resources (water-use payments were introduced in 1981) and completed by the introduction of pollution charges in 1991. These measures were introduced at a time when the centralized planning system was falling apart, in the hope that they would be able to provide a workable alternative to failing command-and-control methods.

The way in which the 'economic incentives' were introduced, however, was closely tied to the structure of enterprises and state planning. For pollution charges an example of this linkage was that a 'standard' charge was levied on emissions and was considered a part of enterprises' normal costs, followed by a higher charge (five times the standard rate) levied on what planners considered to be excessive pollution; this charge was payable out of enterprises' 'surplus income' and has continued to be assessed for emissions above permitted standards.

In the former system of economic regulation, the director of an enterprise was appointed from above and was given so-called technological blueprints by his or her ministry. These plans included calculations of payments for emissions up to levels expected for enterprises using the existing technologies efficiently. In its allocations to enterprises, ministries transferred funds to enterprises to cover these costs. Payments above these levels were recognized to be the responsibility of enterprises and were payable from surplus income, although in the absence of a right to use surplus income freely, this was a rather artificial procedure. The state also made frequent exceptions to the charges; for example, all agricultural enterprises were released from water-use payments for irrigation.

This system and the successor currently in place rely on maximum permissible levels of pollution (MPLs), which have been developed for each enterprise. The rules for these are set out for air, water and soil pollution, and solid-waste disposal, and are described in federal laws. To illustrate how they work, we consider the case of air emissions. For air pollution the main law is the Law of the Russian Federation On Atmospheric Air Protection, passed in 1982. Under this law, MPLs were calculated for each enterprise and local area in each Russian province, using approved computer programs and set at levels which guaranteed that emissions from all pollution sources would not surpass maximum allowable concentrations (MACs) for each substance. MACs are essentially ambient air quality standards which define the allowable concentrations in the atmosphere which supposedly causes no immediate harm and also does not lead to long-term negative effects on human health. Managers of enterprises are obliged to inform the various organizations responsible for air quality monitoring when emissions form concentrations exceeding MACs several times, and are therefore defined as emergency situations.[1] In unfavourable meteorological conditions, after receiving notice from local environmental authorities, polluting enterprises must change their production processes.[2]

A pollution permit is a contract between a polluter and the regional environmental authority, as well as the local department of the State Committee for Sanitary and Epidemiological Inspection. MPLs constitute the basic standard in permits, which are the regulatory basis for the payment of charges. It should be noted that enterprises had nothing to gain from reducing pollution below MPLs, and before the breakup of the Soviet Union even the incentive to keep to that norm applied only to the extent that enterprises cared about loss of surplus income or were unable to persuade authorities to waive excess charges.

On the damages side, we can ask whether the structure was based on strong scientific evidence of acceptable levels of emissions. There is no

---

1   The so-called especially dangerous level of pollution starts at 50 times the MAC level.
2   For example, in Kemerovskaya province in West Siberia, where air turbulence is weak, there are four regimes of industrial operation.

clear answer to this question, but we do know from recent health risk studies that there are no evident thresholds that can be applied to most pollutants. Therefore, the setting of MAC levels, even if they were achieved, probably did not represent optimal control strategies (Markandya, 1994).

This chapter focuses only on pollution charges and does not deal in any detail with other economic instruments that affect environmental pollution. It is worth noting, however, that there are a few other economic instruments. One is the fuel excise tax, which amounts to 3–5 per cent of retail fuel prices. The tax level is dependent on the sulphur content of fuels and this provides some incentive to shift to lower sulphur fuels, although the magnitude of incentive effects have not been analysed. Revenues from the tax are intended to be earmarked to a special fund, similar to the environmental fund, but as of 1995 it had not been established. Another positive environmental incentive is provided by a tax exemption for enterprises providing environmental services as part of an officially adopted programme of environmental action. Participating enterprises receive a 25 per cent reduction in profit taxes.

The remainder of this chapter is organized as follows: Section 2 looks at the implementation experience with pollution charges in Russia, including the basis on which charges are set, how the transition issues are handled and how data are collected. Section 3 evaluates the impacts of charges and Section 4 looks at possible reforms.

2   IMPLEMENTATION EXPERIENCE: SETTING CHARGE RATES, MONITORING AND ENFORCEMENT

2.1   The Basis for Setting Charge Rates

Three basic concepts were considered during an experimental introduction of pollution charges in selected Russian provinces and cities in 1990. This accords with the practice at the time in the Soviet Union where different methodologies were tried in regions on an experimental basis before a methodology was used nationally. For pollution charges the experiment lasted one year and a single approach was adopted in 1991. Damages caused by pollution, costs of pollution abatement and revenue needs were considered as bases for charge rates in Russia, but only damage and revenue-based charge approaches were actually tried. The use of abatement costs to calibrate charges was only considered on a theoretical level (Danilov-Danilyan and Kozeltsev, 1990).

The damage approach bases charges on the effects of pollution on the economy and human health, with a view to compensating the affected parties, and one can find several references to this function of pollution charge in official documents. There were several reasons why the damage method proved inadequate for the calculation of pollution charges. First, the existing methods of estimating damages were relevant to a centralized economy, but did not correspond to market-based methods which would have been more appropriate to the transition conditions in which the charges were being levied. Moreover, with the strengthening of court and arbitration systems, people could challenge the basis of calculating these damages and therefore the importance of publicly-organized compensation of damages was reduced.

Second, the level of aggregation of data was too high and the results of calculations were therefore very approximate; this meant that the damages from particular actions could not be verified. The health damage from pollution was also estimated on the basis of losses in production rather than loss of welfare. The former, of course, underestimates the latter.

Finally, the lack of environmental audits did not allow one to demarcate between past and current damages, and this led to overestimations of damages, which would have imposed excessive burdens on enterprises. The methodology developed to estimate damages also employed a mix of different indicators and it confused stock and flow parameters.

Although the damage approach had many problems, the structure of charges finally adopted did retain some of its aspects. Emissions were differentiated by toxicity, and payment per ton of each substance was established proportional to its potential danger to human health.[3] The gap between the most and least toxic substances was several orders of magnitude; this difference was considered by the designers of pollution charges to provide enough motivation for managers of enterprises to cut levels of the most hazardous pollutants.

Damages were also included by levying a fivefold penalty for all emissions exceeding permitted levels of pollution. Regional characteristics and peculiar territorial conditions were taken into account by introducing so-called 'coefficients of ecological conditions' which varied according to region and ecological circumstance.

The second approach would have based pollution charges on the marginal cost of reducing pollutant emissions (MAbC), with the MAbC measured at the desired level of emissions. With a charge of that level,

3   In most cases a standard four-group classification used by sanitary inspectors was used, with a coefficient equal to the inverse value of the MAC providing the toxicity weight.

polluters have an incentive to cut emissions to the desired level because up to that point the savings in charges exceed the abatement costs.

It is hard to say much against this idea, except to note that incentives only work if enterprises can benefit from reducing their pollution charge payments. Under the centrally-planned system enterprises did not benefit, and as of 1990 all companies were still state owned and operating under soft-budget constraints. In any case, because of a lack of information and skills the proper estimates of costs were never made and this approach was never implemented.

The third rationale for pollution charges was to raise funds. The fiscal system was collapsing in 1990 and funds for environmental protection were becoming extremely scarce. The system of pollution charges, it was believed, could potentially provide extrabudgetary funds for environmental protection. In Russia this was termed the 'tax approach' to environmental charges (Danilov-Danilyan and Kozeltsev, 1990).

At that point in time some experience with earmarked charges already existed. In Estonia, for example, a regional environmental fund had been operating effectively on a demonstration basis. Based at least partially on this successful experience, the tax approach, including the allocation of financial resources to extrabudgetary accounts, was approved as the conceptual basis for the system of pollution charges in Russia.

2.2 Description of the Charge System for Air Pollution in Russia

The methodology adopted for calculating pollution charges is based on the estimated average cost of implementing programmes to reduce emissions. In 1991 the state estimated an annual environmental programme for three media: air, water and solid wastes. If, for example, the total cost of the air programme is C rubles and estimated total annual air emissions (with different emissions weighted by toxicity) amounted to M tons annually, the charge is then computed as the average cost, p, where:

$$p = C/M.$$

This calculation is adapted to each region by multiplying p by the region-specific 'coefficient of ecological conditions' mentioned above. The result is a system of charges for different pollutants, some of which are listed in Table 6.1. The basic value of p in 1991 for atmospheric pollutants was 3.3 rubles per ton of adjusted emissions; for water 443.5 rubles per ton of adjusted discharges. In total there are charges for 240 air emissions, but about 80 per cent of total revenues come from nitrogen oxides and carbon monoxide.

Table 6.1   Charges for air pollutants in Russia in 1995

Pollutant	Allowed emissions (rubles/ton)	Allowed emissions ($/ton)	Above limit emissions charged at 5 times base rate (rubles/ton)	Above limit emissions charged at 5 times base rate ($/ton)
Sulphur dioxide	5,610	1.22	28,050	6.10
Carbon monoxide	85	0.02	425	0.09
Lead	935,000	0.20	4,675,000	1,016.30
Dust (non-toxic)	1,870	0.41	9,350	2.03
Nitrogen oxides	4,675	1.02	23,375	5.08
Benzopyrene	280,500,000	60,978.26	1,402.5m	304,891.30

Note: 4,600 rubles = US$1.

As noted above, pollution charges are differentiated by region. In practice, this differentiation depends on *both* the state of the regional economy and the state of environmental quality. For example, air pollution coefficients vary from 1.0 in the Russian Far East to 2.0 in the Ural region.

This methodology has several problems, of which perhaps the biggest is the choice of '*C*.' In practice, *C* is determined based on past years' actual expenditures and on the basis of recognized needs and priorities. Another problem is indexation for inflation. Though charge rates are indexed (for example, in 1996 rates will approximately double compared with 1995), they are indexed based only on estimated changes in costs of environmental research and investments, and changes in the minimum wage. These costs have lagged behind the overall economy and real rates have therefore declined over time.

2.3   The Charge System During the Economic Transition and the Use of TCLs

The introduction of any new set of policies requires a transition period for a smooth integration into the existing system. A system was therefore developed which defined temporary compliance levels of pollution (TCLs), which were agreed between polluters and regional environmental authorities. The TCLs become pollution permits which increase in stringency to reach MPLs during a fixed period of time (for example, five to ten years).

A TCL is set above the true standard (that is, the MPL). To provide incentives to meet MPLs, emissions less than the TCL but greater than the MPL are treated as beyond firm-level standards, and therefore enterprises must pay a fivefold penalty on those emissions. If a firm exceeds its TCL, it is charged a 25-fold penalty.

Payments of pollution charges for emissions below MPLs or TCLs are treated as part of operating costs. For state enterprises in the centrally-planned economy, this was important because the state covered such costs, but did not cover charges on emissions above permitted levels. As enterprises are privatized, the distinction between operating costs and profits matters only to the extent that enterprises actually make taxable profits.

TCLs are supposed to be determined based on the level of emissions expected from similar enterprises using best-available technologies (Ministry of Construction, 1990). The definition of the best-available technology is, of course, problematic, and in practice a TCL is the result of a negotiation process which considers many factors, including the financial state of enterprises, the costs of environmental actions and the sizes of enterprises. Enough experience has, perhaps, now been collected for this *ad hoc* procedure to be replaced by a more formal mechanism for establishing TCLs. This remains to be done.

2.4 How MPLs for Air Pollutants are Determined

The basic standard of environmental quality for monitoring and enforcement is the maximum ambient concentration (MAC). MACs in the atmosphere are the maximum concentrations of substances, averaged over various periods of time, which are deemed not to cause any hazardous impact on human health (including long-term effects) or the environment. As noted earlier, this notion is itself suspect.

MACs are converted into maximum permitted levels of emissions (MPLs) for each enterprise by modelling the links between emissions and concentrations, allowing for existing background concentrations. MPLs are therefore the amount of a hazardous substance discharged into the atmosphere, taking into account emissions from all other sources, projected development in the region and dispersion in the atmosphere, which does not generate concentrations exceeding MACs.

An MPL is developed for each source of pollution, and is a rather complex procedure which relies heavily on polluting enterprises themselves. The first stage of this process is the construction of an inventory of sources of pollution in a region. Each enterprise prepares its own inventory. In doing so, it typically relies on indirect (balance) methods of calculation rather than on direct measurement. The second stage is the creation of a so-called technological passport of emissions, including a description of each source of pollution, the purification equipment installed, the height of stacks, and the temperatures, speeds and volumes of emissions per stack. As with the first stage, each enterprise provides this information. Because, according to the legislation, MPLs should be calculated under actual conditions of operation, in the third stage detailed information is provided on the equipment to be used (for example, age, quantity and quality of fuel, and so on).

The fourth stage is the calculation of existing emissions in tons per year. This is done by specialists from the lead branch institute of the sector of the economy to which a polluting enterprise belongs, or by laboratories located in large enterprises. Stage 5 includes estimates of ambient air quality concentrations, also made by the lead branch institute in coordination with the lead regional institute, which is usually part of the Ministry of Environment. These calculations are made using mathematical models called 'Zephyr 6' and 'Warrant' developed by the former Soviet State Hydrometeorological Committee. If these models estimate concentrations exceeding an MAC, then a list of activities to diminish emissions is prepared by each enterprise and the procedure is repeated with new levels of emissions. Iterations continue until the concentrations for each substance are less than or equal to the MAC. The final product of the model is an MPL for each pollutant and each source of pollution in each enterprise.

The pollution permit itself consists of two parts: a compliance requirement to meet certain pollution levels and an appendix. The appendix includes a list of polluting substances, actual emissions and an MPL or TCL for each pollutant emitted at each source. When a source is assigned a TCL, a list of actions required to comply with the MPL is included along with expected costs to reach the MPL. The pollution permit may be cancelled if abatement measures are not taken in time, and under such circumstances all emissions are treated as above-limit and the 25-fold penalty is levied.

Enforcement of permit terms pertaining to emissions into the atmosphere and water is the responsibility of Regional Environment Committees of the Ministry of Environment. Monitoring, however, is primarily in the hands of enterprises which provide Regional Environmental Committees with reports on pollution every trimester.

2.5   Rate of Profit Constraints on Pollution Charges

As we have noted, the system of pollution charges in Russia during the transition period involves a series of *ad hoc* compromises between environmental and economic objectives. One area of compromise is in the

Table 6.2　Maximum percentage of profit that pollution charges can constitute

	Profitability (in %)		
	up to 25	up to 50	more than 50
Maximum pollution charges as a percentage of total profits	20	50	70

setting of TCLs. Another is with regard to the pollution charge payments actually made by enterprises.

In order to avoid levying charges which exceeded company profits, upper bounds on pollution charge liabilities have been established. These maxima are based on the profitabilities of enterprises, which apply the regulations and calculate for themselves the maximum percentage of their profit that pollution charges can constitute.[4] These calculations are then checked by regional environmental committees and approved by local authorities.[5] Table 6.2 reports these maximum values.

This constraint is binding for many enterprises and indeed all enterprises which make zero profits pay no pollution charges. Although there are no confirmed data available on how this rate-of-profit constraint affects total charge collections, it is known that in 1995 the Ministry of Environment collected only half of the 600 billion rubles ($130 million) it anticipated. Estimates are that about two-thirds of this gap can be explained by the lack of profits reported by enterprises, combined with poor enforcement.

Capping pollution charges in this way is formally considered a privilege and is therefore given only to the following sectors of the economy (Ministry of Environment, 1993):

- energy sector, including fuels and oil refining;
- agricultural products and medicines;
- conversion of military production;
- transport and communications;
- housing construction.

4　Profitability is calculated as total profit divided by total direct costs.
5　For example, without some sort of maximum, pollution charges would constitute 153 per cent of profits in the energy sector and 253 per cent of chemical industry profits.

In addition, organizations engaged in scientific activities or works related to defined 'urgent social needs' have access to this cap on pollution charges. As with other compromises, this method is an *ad hoc* way of dealing with the transition, overlaid on to the system of TCLs and other mechanisms that are designed to address the same issues. The objectives of economic restructuring and pollution control are both important, but reconciling the conflicts between them could potentially be done using a more integrated approach.

3   EVALUATING THE SYSTEM: EFFECTIVENESS OF THE POLLUTION CHARGE AND PERMITTING PROCESSES

3.1   Revenues from Environmental Charges

The revenues from pollution charges are transferred to the Federal Environmental Fund and to regional environmental funds established in 1992. Approximately 80–85 per cent of the revenues of the system of environmental funds come from pollution charges (Gofman and Golub, 1995).[6] The Federal Environmental Fund receives a relatively small part (less than 5 per cent) of the revenues provided to the environmental fund system, and this minority position is considered an important component of the process of decentralization of environmental policy. The major reason why the Federal Fund is so small, however, is that non-payment of charges is quite common. This feature is very general in the Russian economy and is related to problems of economic inefficiency and weak enforcement of bankruptcy procedures.

The overall level of environmental expenditures in Russia in 1994 was approximately \$2 billion, of which capital investments made up \$592 million. The total amount of extrabudgetary funds (generated primarily from pollution charges) for environmental investments is very small and totalled only \$14 million in 1994, or 2.4 per cent of total environmental investments. Most environmental investments are, indeed, financed out of the funds of joint-stock companies (about 51 per cent) or state and municipal enterprises.

The proportion of Federal Environmental Fund resources spent on actual investments is substantially higher than for the system as a whole. The share of environmental investments by the Federal Fund as a

6   Other sources are environmental damage compensation, charges for destruction or improper use of natural resources, fines for infringement of environmental legislation and voluntary contributions.

Table 6.3 *Distributions of the Federal Environmental Fund (in aggre-gate and as a percentage of total fund expenditures)*

Type of environmental activity	1992	1993	1994
Financing of waste utilization, monitoring and resource-saving technologies	0.1 (54.1%)	1.5 (75.3%)	3.1 (78.0%)
Other activities	0.085 (45.9%)	0.49 (24.7%)	0.50 (12.7%)
Total	0.185 (100%)	1.99 (100%)	3.97 (100%)

Note: Millions of US dollars in 1994 prices.
Source: Gofman and Golub (1995).

percentage of its total expenditures was 54 per cent in 1992, the year when it was established, and 78 per cent in 1994. Table 6.3 gives the allocations from the Federal Environmental Fund for the years 1992–94. Projects financed in 1994 included the construction of a plant for electroplating slag utilization and ceramic production in Voronez AO 'EPROM' ($0.58 million); installation of equipment for chemical sludge reprocessing and more efficient cement production in Volgograd ($0.55 million); and completion of a plant for chemical sludge utilization in Balakovo ($65,217).

3.2 Evaluating the Overall System of MACs, MPLs and TCLs

A review of the permitting procedure leads to the conclusion that there are many problems and there is significant room for improvement. The first problem is the complexity of the process and the substantial controls placed on each enterprise. The authorities need to collect as much information, and to monitor enterprises as much as in any command-and-control system. Administration costs are therefore high and enforcement is poor.

Second, emissions measurements are based mostly on indirect methods – using a materials balance approach. This provides only rough estimates of the actual emissions, as the functioning of plant and equipment is generally not in accordance with technological assumptions. The materials balance approach is employed based on the assumption that equipment is operating in an ideal way. In general this is not correct, and estimates suggest that about one-third of total beyond-limit pollution occurs because of poor operation of technological processes and purification treatment (Kozeltsev, 1993).

Although the total number of charged substances is quite large, and in all official documents relating to emissions control the importance of direct measurement is proclaimed, there are only a few pollutants for which actual measurements are made. These contaminants are dust, CO, NO_x and SO_2. In cities and towns which are under conditions of ecological crisis, however, additional substances are measured.[7] Inspectors directly control approximately 29 water pollutants and 44 air pollutants in such cities (Ministry of Construction, 1990). Expanding direct measurement will significantly improve the system of charges by verifying levels of emissions at sources of pollution.

Related to the issue of measurement is the problem of poor and incomplete recording of emissions. A large number of sources are indeed yet to be inventoried. Among them are nonpoint sources and sources located in small and medium-sized enterprises. Estimates from the Moscow State Committee on the Environment indicate that in Moscow about 40 per cent of pollution sources are not reflected in the inventory (Kozeltsev, 1993).

Because enterprises themselves determine their emissions, understating emissions is a frequent phenomena, and recording MPLs as actual emission levels is also quite common. Laboratories measuring emissions exist only in large enterprises, and in medium and small enterprises this is a duty of production units, whose employees may not be adequately trained.

Conflicts between allowing economic activities to continue without interference and meeting environmental standards are generally resolved in favour of the economy, and permits for new enterprises setting up in an area often include generous TCLs. Despite the requirement that TCLs be set based on best-available technologies, TCLs are often simply set equal to existing emissions levels. Where authorities feel compelled to take action to reduce ambient air concentrations, they often do this by allowing enterprises to raise stack heights and pass on environmental damage to neighbouring areas. While it may be correct to allow economic activity to continue and not to impede the revival of enterprises with overly strict environmental regulations, the *method* by which such decisions are taken is not one that will result in an efficient allocation of resources.

It should also be noted that there are some doubts regarding the dispersion models used. Valid information on prospective regional devel-

7   More than 100 places in Russia have this official status.

opment simply does not exist, and this lack of information forms an important block to the use of the computer models. Perhaps more fundamentally, however, doubts also exist regarding several features of the computer models themselves.

3.3   Evidence on Impacts of the Pollution Charge System

A review of studies of the Russian experience with the use of the pollution charges since 1990 shows that the introduction of charges has not resulted in significant increases in environmental investments or reduced emissions. In one study of pollution-abatement costs and pollution charges, for example, engineering estimates of abatement costs in most industries exceeded charge levels by 2.3 to 2.9 times (Golub and Kolstad, 1995).

Another study established a correlation between the low pollution charge rates and their moderate impacts on polluters. Pollution charges levied in Moscow Municipality were estimated to influence only 8 per cent of enterprises generating above-permitted levels of air and water pollution. In order to have a more significant impact on Moscow polluters, it was estimated that charges would have to be quadrupled (Kozeltsev, 1993).

The same study also examined the range of pollution charges within which companies would be expected to respond by reducing emissions. Directors of the most polluting enterprises in five Russian cities were surveyed and it was found that there exists a so-called 'sensitive area' for managers in which pollution charges reduce their profits not less than 5 per cent, but not more than 20 per cent. Directors reported they believed that when pollution charges exceeded 20 per cent of profits, they would not in reality be required to pay the charges.

If pollution charges have not achieved reductions in emissions, what did they achieve? We believe there were some positive impacts. First, they represented the start of the development of an emissions inventory and the beginning of increased consciousness about pollution goals. Because it provided at least some impetus for the Russian business sector to develop monitoring and improved audit systems, enterprises are now looking – at least somewhat – at possibilities for cost savings from more efficient operations. Related to that, the process of getting permits from environmental authorities requires the formulation of a medium-term strategy for reducing pollution. This strategy gives an incentive to at least list the necessary steps to reduce emissions.

Assisted to a small degree by the system of pollution charges, the process of economic reforms stimulated a search for new low-cost environmental options. Increased competition among manufacturers, the strengthening of the small business sector and the development of a service sector has led to a more stable demand for low-cost cleanup facilities.[8] New, compact cleanup equipment is better suited to local conditions.

The charges have also provided one of the few sources of finance for environmental monitoring and investments at the federal, state and regional levels. As was stated in Section 2, the need to create a new system for environmental finance distinct from the normal state budget system was one of the main arguments for the introduction of pollution charges.

4 CONCLUSIONS AND MAJOR ISSUES TO BE ADDRESSED

The system of pollution charges in Russia is imperfect, inadequate and in need of extensive reform. In the present economic circumstances, however, and given the government's need to deal with perhaps more pressing issues, it is unreasonable to expect a major overhaul of the system in the short (for example, 2–3 year) term. Our proposals for improving implementation and effectiveness fall, therefore, into two categories: short-term steps and medium- to long-term measures that can be initiated now, but implemented over the period 1996–2005.

An important short-term issue to be addressed is the haphazard and *ad hoc* nature of measures adopted to ease the transition. Enterprises can negotiate away charges effectively, given loopholes such as the availability of TCLs and charge limits based on profitabilities of enterprises. Where there are conflicts between meeting environmental goals and meeting economic objectives, these should be addressed using risk-assessment techniques rather than informal methods.

In the short term there is also likely to be an important role to be played by charges as sources of revenues for environmental expenditures, and it is unlikely that they can be replaced by anything better in the near future. It is therefore important to ensure that revenues are maintained in real terms with better indexation. As economic activity picks up, charge levels could in fact increase *faster* than inflation. A comparison with real charge levels in Central and Eastern European countries could potentially provide an indication of what is practicable.

A complex but important issue with regard to the charge system is the operation of the system of environmental funds which draws its revenues from the charges. Presently, there are several levels of funds and decisions

8   This is in contrast with the situation in the former Soviet Union, where a few monopolistic enterprises produced expensive cleanup equipment to service the dozens of plants located near big cities.

on which expenditures to fund are taken without proper appraisal. Enterprises can also be exempted from paying charges if they make environmental investments, but there is no analysis on whether such investments are the most effective use of those funds.

In the medium and long term, the question of whether the Russian Federation should go for pollution charges *at all* needs to be analysed. The present system is very information and data intensive, and requires a sophisticated system of monitoring and enforcement even though there is virtually no continuous emission monitoring anywhere in Russia.

Under such circumstances it is possible that taxes on key inputs (fuels, chemicals, and so on) levied at the point of manufacture or distribution would be much less costly from an administrative viewpoint. They could also be set in *ad valorem* terms, avoiding the problem of indexation.

Input charges are also not without their problems, however. For example, such charges are less closely related to marginal damages or abatement costs than pollution charges.[9] Input charges also remove incentives for end-of-pipe cleanup.[10] Restoring such incentives must therefore be accomplished through rebates to enterprises making such investments. Regional differentiation, which is a feature of the pollution charge system, may also be a problem with input charges. Given these disadvantages, it is unlikely that they can replace all pollution charges, and a few charges on key emissions, combined with direct regulations, will probably still be desired.[11]

The final issue is the weakness of institutions that implement environmental regulations in Russia. They are grossly underfunded and the demarcation of responsibilities between different agencies and between federal, regional and local authorities is not always clear. These vagaries will have to be cleared up if environmental policy is to improve in the future.

REFERENCES

Danilov-Danilyan, V. and M. Kozeltsev (1990), 'Vybrosy za platu' (Charges for Pollution) *Voprosyi Economici* (Economic Problems), **1**: 120–30.
Gofman, K. and A. Golub (1995), 'The system of environmental funds in the Russian Federation', Paper presented at the United Nations Workshop on

9   It is, of course, also true that the present system is largely unrelated to these variables.
10   For example, if you pay a tax based on the sulphur content of coal, there is no incentive to invest in desulphurization equipment.
11   Indeed, for some environmental problems command-and-control instruments are still the most effective.

Economic Instruments for Sustainable Development, Pruhonice, Czech Republic, January.

Golub, A. and C. Kolstad (1995), 'The use of economic incentives for the control of pollution in Russia', Mimeo.

Kozeltsev, M. (1993), 'Nalogi i stimuli' (Taxes and Incentives), *Zelenyi Mir* (Green World), **1**: 4.

Markandya, A. (1994), 'Externalities of Fuel Cycles: External Project Economic Valuation', Document No. 9, European Commission, DGXII, Brussels.

Ministry of Construction (1990), 'Prirodoohranyi normi e pravila proektirovanya spravochnik' (Environmental Project Norms and Rules Handbook), Moscow.

Ministry of Environment of the Russian Federation (1993), 'Instruktivno metodicheskie ukazania po vzimaniu platy za zagryaznenie okruzayushei prirodnyi sredi' (Guidelines for Levying Pollution Charges), Moscow.

Ministry of Environment of the Russian Federation (1994), 'Ekonomika prirodopolzovania: Analyticheskie i normativnye materialy' (Environmental Economics: Analytical and Normative Material), Moscow.

7. Integration of Pollution Charge Systems with Strict Performance Standards: The Experience of the Czech Republic

Zdenek Stepanek

1 INTRODUCTION

Thanks to political changes which occurred in 1989–90, it was possible for the Czech Republic to revise its approach to environment protection and to formulate a more efficient environmental policy. Qualitatively new basic legislation was therefore developed during the years 1991–93. The new legislation was based both on amendments to the existing laws, as well as on the formation of completely new legal standards.

New legislation was inspired mainly by the legal standards of the European Union and is focused on administrative instruments of environmental policy (limits, technical standards, and so on). But this structure is also complemented by the use of economic instruments. Pollution charges and subsidies (primarily grants and soft loans) are the most significant examples.

Pollution charges were gradually introduced in the Czech Republic in the 1960s. Air pollution charges were introduced in 1967, and wastewater charges for disposal of effluents into surface waters were instituted in 1979. But in a centrally-controlled economy they played no significant role as their effect was completely devalued by the deformed economic relations. Since 1991 these charge systems were, therefore, reformulated based on completely new principles.

The current system of environmental charges includes:

- air pollution charges;
- water pollution charges;
- solid-waste disposal charges;

- charges for the offtake of water from waterways;
- charges for the withdrawal of underground water;
- levies for the sequestration of agricultural land from the agricultural domain;
- charges for the mining of minerals.

With regard to the first three charge systems, the stated goals are first to encourage polluters to implement measures leading to decreased pollution emissions. The systems should also contribute to providing financing for environment protection and also compensate for damages to the environment. The charges also play an important role in the transition to a market economy, because they equalize the economic conditions between individual polluters. Without charges, an environmentally-friendly enterprise is put at a disadvantage, because its costs will be higher. It will therefore only with difficulty compete with more polluting enterprises.

With the help of charges, therefore, environmental externalities are at least partially internalized into the costs of polluters. Various tax breaks complement the system of charges. These include, environmentally-motivated income tax releases for investors in alternative sources of electricity and a reduced value-added tax (VAT) rate from 22 to 5 per cent for ecologically-friendly products. Unfortunately, energy from power and heating plants burning environmentally-non-friendly brown coal is also taxed at this reduced rate.

The system of air pollution charges is a typical system in which we can illustrate the advantages and disadvantages of contemporary charge systems. This charge system is also the most important one, because the effort to improve air quality is the number one environmental priority. The next sections will therefore be dedicated just to problems associated with air pollution charges.

Section 2 describes the air pollution charge system in the Czech Republic and discusses its most important features. Section 3 deals with the lesson learned from this system and Section 4 attempts to evaluate the efficacy of the charge system. Section 5 discusses the main problems to be solved together with possible solutions and Section 6 summarizes the main conclusions.

2 DESCRIPTION OF THE AIR POLLUTION CHARGE SYSTEM

Air pollution charges were introduced in the Czech Republic in 1967. This law was valid without changes until 1991 and therefore was outdated. It was therefore necessary to amend not only charge rates, which remained unchanged for 24 years, but also the whole charge system. Revisions included a redefinition of the polluting substances liable to charges, the definition of pollution charge payers and a removal of the link between charge rates and the heights of stacks. In 1991 two new laws and associated regulations were passed which replaced the preceding ones: the Air Protection Act and the Act on the State Administration of Air Protection and Pollution Charges. According to these laws, charges were to be paid by operators of all sources of air pollution with the exception of mobile sources. Charges were also not applied to physical persons operating small sources of pollution (up to 50 kW), as long as these sources were not used for profit-making activities.

Authorities for administering this system were based on the size of pollution sources and are as follows.

- *Large pollution sources (above 5 MW)*   The Czech Environmental Inspection (CEI) monitors emissions and assesses and collects charges. There are approximately 1,500 such sources.
- *Medium-size pollution sources (between 0.2 and 5 MW)*   Local district authorities assess charges and collect them. There are approximately 20,000 such sources.
- *Small pollution sources (up to 0.2 MW)*   Local municipal authorities make decisions regarding charges and also collect them.

The charge rates for large and medium pollution sources is determined according to the Annex to the Act on the State Administration of Air Protection and Charges for Air Pollution (1991). This Annex includes a list of polluting substances (approximately 90) which are subject to charges, as well as charge rates for each unit of pollutants emitted. To simplify administration, differentiated rates were established only for five main pollutants (solid emissions, sulphur dioxide (SO_2), nitrogen oxides (NO_x), carbon monoxide (CO) and hydrocarbons (C_xH_y)). All other pollutants were divided according to their toxicity into three classes. Uniform rates were established for these classes as shown in Table 7.1.

As these rates are higher than those in the 1967 legislation, charges were phased in to final levels over a five-year period. This means that the charge actually paid until 1997 is only a fraction of the final rate. The schedule in the legislation is as follows:

30 per cent of final rate in 1992 and 1993;
60 per cent in 1994 and 1995;
80 per cent in 1996;
100 per cent in 1997 and subsequent years.

Table 7.1   *Air pollution charge rates for main pollutants and the three toxicity categories*

Pollutant	Charge rates (Czech crowns per ton)
Solid emissions	3,000
Sulphur dioxide	800
Nitrogen oxides	800
Carbon monoxide	600
Hydrocarbons	2,000
Other:	
Class I	20,000
Class II	10,000
Class III	1,000

Note: 1 US dollar = 27 Czech crowns.

The above schedule is applied only to large and medium pollution sources. When polluters do not meet their emission limits for a given year, the pollution charge rate is increased by 50 per cent for all above-limit emissions. Emission limits in tons are set by the CEI. For small pollution sources, which are liable to charges by municipal authorities, charges are calculated as a fixed amount up to a maximum of 40,000 Czech crowns. The amount of the pollution charge is determined based on the size of the pollution source and the harmfulness of the pollutant.

Pollution charges collected from large sources by the CEI are turned over to the State Environmental Fund (SEF). Charges from medium sources are collected by district offices and are also turned over to the SEF. Small pollution sources' charges are collected by municipal authorities and become a part of municipal budgets.

The Act on Charges for Air Pollution was also the first law into which positive stimuli were introduced. Polluters who make investments leading to emissions reductions pay only 40 per cent of their charges for the whole period of the development of projects.

3   EXPERIENCE WITH IMPLEMENTING AIR POLLUTION CHARGES IN THE CZECH REPUBLIC

As was already stated, administrative instruments are the basis of the air protection system. Source-level emission limits play the dominant role within this class of instruments and are set for all important polluters. These 'new source' limits are in force as of 1992, and are determined based on levels 'which can be attained only with the best-available, reasonably-priced technologies'. More or less, these standards correspond to limits valid in the countries of the European Union. The model for their determination was the German legislation TA LUFT 86.

Emission limits for existing pollution sources are determined by the CEI individually for each polluter, and are based on estimates of the minimum attainable emissions possible from properly operated existing technologies. These limits are valid until the end of 1998 at the latest. Compared with the emission limits for new sources, they are several times less stringent. If by 1998 a polluter does not comply with its emission limit, the CEI will have the power to forbid further operation of the pollution source.

It is supposed that the attainment of emission limits for new sources, which all polluters will be obliged to fulfil after 1998, will ensure acceptable ambient air quality everywhere in the Czech Republic for almost the whole year. There will be some exceptions in a few regions of the Czech Republic which are predisposed to inversions during the winter, but all international treaties dealing with air pollution will be fulfilled. Fulfilling these limits will also imply that 1999 sulphur dioxide, nitrous oxide and dust emissions will be only 40- 60- and 35 per cent, respectively, of 1990 levels.

The permit term and associated emission limits are purely command-and-control instruments. They therefore contain all the advantages and shortcomings of administrative management. This system is not combined with economic instruments in order to offer polluters a choice of abatement levels. Decisive for the behaviour of a polluter is the likelihood of meeting its emission limit by 1998. Economic calculations regarding whether it is more advantageous to pollute and pay charges than to implement abatement measures, are simply not relevant before 1998.

The system, drafted as it was by lawyers and technicians based on standards of countries from the European Union, especially Germany, *a priori* substantially reduced the significance of the pollution charge system. As a consequence, charge rates were set at substantially lower than optimal levels. Their role is – in contradiction to the original aim – mostly fiscal. Their influence is also considerably weakened by the phase-in schedule which was introduced because of fear that a sharp increase in charges might meet with resistance from polluters. As of 1992, for a majority of polluters, charges represented less than 1 per cent of production costs. Their stimulating effect was therefore relatively small and the situation is unlikely to have changed by 1995 (Federal Committee of

Table 7.2   *Incomes and expenses of the State Environmental Fund*
           (millions of Czech crowns)

	Total air pollution charges	Total SEF revenue	SEF expenses for air protection	SEF total expenditures
1991	66	1,592	133	1,537
1992	783	2,412	510	1,492
1993	819	2,768	937	2,895
1994	1,415	4,489	1,062	3,584

Note: 1 US dollar = 27 crowns.

Environmental Protection, 1992a). From a fiscal standpoint, however, as shown in Table 7.2, charges represent a non-negligible part of the SEF revenue stream. Most air pollution charges are used to subsidize projects aimed at improvement of air quality.

As was already noted, emissions limits for existing sources, as well as the allowed term of their operation are determined by the CEI. The polluters themselves inform the CEI about the volume of emissions once per year using a methodology developed by the CEI. The CEI also carries out random controls of the accuracy of these calculations. Only the most important polluters must install equipment for continuous automated measurement of emissions by the end of 1997, but this will cover approximately 70 per cent of all large sources.

4 EVALUATION OF THE CHARGE SYSTEM

Although the stated primary goal of pollution charges is to stimulate pollution abatement, the charge rates set in 1991 do not fulfil this function. Already in the year of their introduction the rates determined for the so-called main substances deviated from values even roughly corresponding to abatement costs. This point was particularly evident in the case of the charge rates of $30.00 per ton for SO_2 and NO_x emissions. Estimated charge rates to stimulate significant investments in abatement equipment are $240.00 and $277.00 (VUPEC Economy, 1995).

Moreover, indexation was not embodied into the law and therefore pollution charge rates were not adjusted to inflation (see Table 7.3). Annual inflation rates were quite high, however, and as a result the

Table 7.3   Annual inflation rates in the Czech Republic, 1991–1994

Year	Inflation rate (%)
1991	57
1992	11
1993	20
1994	10
1995	9

deviation from abatement costs increased over time. To correspond at all with abatement costs, the final level of charges, which will be valid only in 1997, should be approximately ten times higher for SO_2 and twelve times higher for NO_x emissions. The amount of charges has also been depreciated by inflation.

At the same time there should, of course, also be provisions for delays and relief from pollution charges when polluters undertake projects to reduce emissions. The 40 per cent figure for the amount of charges paid by polluters during the development of an environmental investment project was accepted as a political compromise, closely linked to discussions on the degree of decentralization of charge payments. It was not set as a result of economic calculations.

The decomposition of the phase-in period into five one-year increments was also not reasonable. Even if the pollution charge rates were determined correctly, they only begin to fulfil their function in 1997. Up to that time it will be more advantageous for polluters to pay pollution charges than to build and operate purification equipment, or to take other steps to protect the environment. Starting in 1999, existing sources will have to comply with the even stricter limits currently only applicable to new pollution sources, and therefore the current system of pollution charges will really only fulfil their function during two years (1997 and 1998).

A thorough evaluation of the charge system in the area of air protection is really possible only in the context of the whole system of air protection. Unintended economic impacts of the tightening of standards which will occur in 1999 provide one set of examples. What are the choices open to polluters as they approach 1998? Polluters really have only two possibilities. Either they adhere to the emission limits for each source regardless of costs or they must liquidate all noncompliant sources. This strict approach has important efficiency consequences, because sources which in 1998 will be close to the end of their service lives must either be reconstructed to fulfil the 1999 limits or they will have to be put out of operation. This must occur even when they could be operated, for example, as auxiliary or reserve equipment. Sources which in 1998 will still be in relatively good condition (for example, not older than 12 years), but which cannot be reconstructed so they can meet the 1999 limits will also have to be put out of operation. The most affected will be technologies installed shortly before 1991.

Even without considering the changes which will take place in 1999, the system based on standards is certainly less than effective. For example, the existing system of air protection does not at all optimize costs of emissions reductions. Practically every polluter has different costs for reducing pollution, and therefore emissions may be reduced at various costs. The global interest is to minimize the costs of pollution abatement, but the existence of different costs and the possibility of optimization of total costs were not considered.

Monitoring is also a problem. The existing system uses instruments which do not allow early verification of declared emissions. For example, the CEI has no way to track and check the fulfilment by sources of CEI directives. It is therefore not able to intervene early and effectively if sources do not comply with CEI regulations. As a result of this lax enforcement, a non-negligible percentage of polluters are doing nothing to reduce emissions,[1] and are presumably expecting that it will be infeasible for the CEI to close so many economically significant pollution sources. They expect the deadline for meeting emission limits to be pushed back, or that various exceptions will be introduced. As of 1995, approximately 25–30 per cent of polluters were estimated by the CEI to be behaving in this way.

The contemporary system of air protection also does not ensure that polluters have access to necessary financial resources under acceptable conditions. The insufficiency of cost-effective long-term credit represents a significant barrier to the realization of usually costly measures to reduce emissions.

The existence of strict limits for new pollution sources also does not provide polluters with incentives to look for the very best solutions to pollution problems, because they receive no benefit for incurring the higher costs necessary to go beyond their standards. The saving of emissions can neither be sold (for example, to other polluters) nor used in any other way. In this way non-progressive technological solutions are

---

1   Relatively low pollution charges do not help, either.

conserved; although the principle of encouraging the use of best-available technologies is declared, the system does not really support this goal.

An important distributional problem is that until 1998 the charge system favours those who do not reduce emissions. As the level of pollution charges does not correspond to the costs of emission reductions, and will not correspond even after attaining the final levels in 1997, all new polluters and all current polluters who decrease their emissions are put at an economic disadvantage. Paradoxically, the old industrial structures which pollute the most also benefit more from the imperfectly constructed charge system.

5   POSSIBLE SCENARIOS FOR THE FUTURE OF AIR PROTECTION IN THE CZECH REPUBLIC

The negative impacts of the present legislation (especially its high compliance costs) are so grave that they endanger the fulfilment of the system's main goals. There are two broad approaches to revising the system.

1. Improve the system of administrative management by strengthening the existing instruments and at the same time tightening state administration.
2. Replace the existing system, which is based on standards, and fully use the possibilities of economic instruments. These could include an improved charge system or perhaps tradable permits.

Pure forms of either of these two approaches are not feasible. The first one would meet with opposition from the economy and the second one would be opposed by the bureaucracy, which is dominated by lawyers and technocrats who prefer administrative instruments.

Because the 1998 deadline for fulfilment of source-level standards is so critical, but as of 1995 the situation is still unclear, it is perhaps most interesting to ask what direction the environmental policy might take as the 1998 deadline approaches. Two main policy visions are possible.

The first possibility is that the current system is revised based on pressure from special interests. Under this scenario, polluters will obtain delays for their meeting the standards. The economic impacts of the high-cost air protection system will therefore not be removed, but will merely be delayed. As of 1995 proposals had already been made for such delays, and the most frequently proposed solution to the problem of possibly rampant noncompliance was to offer a delay until 2005 for meeting the standards.

Adopting such a programme would lead to a series of undesirable phenomena. These would include a prolonging of the substantial damages believed to be caused by air pollution.[2] It should be noted that most of these externalities are borne by the private sector; environmental costs of noncompliant polluters would therefore continue to be passed on to the population and private businesses. A negative signal about the rule of law and the possibility for exemptions from laws in the Czech Republic would also be sent to potential investors.

Enterprises which produce environmental protection equipment and environmentally-friendly technologies would suffer losses if the introduction of stricter limits was delayed, because the demand for their equipment would be postponed by several years. The system would also continue to favour those who pollute. New businesses, as well as those which, without excuses or delays, accepted emission reduction targets, will continue to operate under a competitive disadvantage.

An alternative policy vision would attempt to reduce the negative economic impacts of the current system and therefore at the same time ensure its feasibility. Such a future is possible, however, only if the flexibility possible from economic instruments is exploited. These instruments would especially focus on pollution charges, but could also be complemented by tax instruments and modified tradable emission rights,.

To achieve this goal, a basic revision of the charge system will be necessary. Charge rates for selected main pollutants (dust, SO_2, NO_x) should be increased to levels roughly corresponding to emission reduction costs. It will also be necessary to ensure that polluters have access to credit at reasonable rates.

Such a step will remove economic differences in costs between polluters and will create incentives for all polluters who find it advantageous not to delay implementation of their emission reduction plans. Polluters who start work on measures to decrease emissions should receive a compliance delay or other relief from charge payments which is greater than the level given in the existing legislation of a 60 per cent reduction in charges. An integrated policy programme of this type would offer polluters who make environmental investments significantly lower charges, while those who do not take the legislation seriously and bet that enforcement will be postponed will be penalized. This charge waiver benefit could even be increased for polluters whose emissions do not exceed the limits valid for new sources and who also make environmental investments. Such high achievers could, in principle, be exempted from

2   Annually, these are estimated to be on the order of tens of billions of Czech crowns (Federal Committee of Environmental Protection, 1992b).

payment of emission charges until the end of 1998. This step would protect Czech industry, because it would have the effect of equalizing the costs of polluters who meet new source limits[3] with those of competitors abroad who do not pay charges.

Why should polluters have incentives to *only* fulfil their limits? Polluters who reduce their emissions below limits determined for new sources should be able to sell these emission differences to other polluters in the administrative district in which the emissions were saved. Buyers of emission rights benefit by increasing their emission limits by the amount of the emission rights bought. Such flexibility would reduce the total costs of emission reductions by adjusting the system to differences in costs of abatement. Individual polluters could therefore minimize their expenses while at the same time total emissions in the region are not increased.

The SEF in its existing form offers mainly grants, and only by a small amount increases the financial resources available for environmental investments. To leverage and therefore increase the effective amount of public funds available, it is desirable to focus mainly on loan guarantees, soft loans, and buy-downs of commercial interest rates.

The producers of electricity and heat who use medium brown coal as a fuel are the most important air polluters. It is therefore very important that electricity and heat, as well as brown coal itself, be reclassified from the 5 per cent VAT category into the standard 22 per cent rate group. The existing reduced VAT rate for brown coal and energy products represents a de facto subsidy. Subsidized energy then leads to wasted energy; at the same time the Czech Republic consumes more than twice the primary energy per unit of GNP than OECD countries (Government of the Czech Republic, 1995).

The removal of this tax advantage for energy is forecasted to result in price increases averaging 17 per cent, leading to conservation and lower energy consumption and consequently a reduction in air pollution. The reduced use of brown coal is expected particularly to impact on emissions of carbon dioxide (CO_2). Indeed, it is anticipated that simply by increasing the VAT on energy in the medium term, it will not be necessary to introduce any charges specifically aimed at carbon emissions.

6 CONCLUSION

Although the system of the environment protection developed in the Czech Republic after 1990 was based primarily on administrative instru-

---

3   Which are comparable with limits valid in countries of the European Union.

ments, we at least in part succeeded in incorporating economic instruments. Among the economic instruments, the most significant are charges. The overall charge system is relatively weak, however, for two main reasons:

- the charge system is unintegrated and is essentially a junior partner to the system of air protection standards;
- the low levels of charges and charge waivers, combined with the long phase-in period, reduced incentives for emissions reductions.

The result is that the charge system fulfils a fiscal function rather than a stimulating one. For polluters it is always more advantageous to pollute and pay than to abate, and this situation will certainly remain unchanged until 1998. An appropriately revised charge system could, however, play a unique role in support of the stricter limits which will be in force starting in 1999. Consideration should also be given to supplementing the charge system with a modified version of tradable emission rights integrated into the system of emission limits. Given the policy inclinations of the existing bureaucracy, however, strengthening and expanding the role of economic instruments in these ways will not be easy.

REFERENCES

Federal Committee of Environmental Protection, Czechoslovakia (1992a), 'Background working paper', Prague.
Federal Committee of Environmental Protection, Czechoslovakia (1992b), 'Report on the State of the Environmental in the Czech Republic', Prague.
Government of the Czech Republic (1995), *Statistical Yearbook of the Czech Republic,* Prague.
VUPEK Economy (1995), Unpublished internal estimates, Prague.

8. Environmental Emission Charges and Air Quality Protection in Hungary: Recent Practice and Future Prospects

Glenn E. Morris, József Tiderenczl and Péter Kovács

1 INTRODUCTION

Environmental emission charges, in the form of fines levied on organizations for exceeding emission limits,[1] predate the 1989 Hungarian 'change in system'. They are part of a mixed environmental protection system that relies on a combination of emission limits and ambient standards to establish performance objectives and permits, fines, administrative requirements, legal sanctions and subsidies to obtain compliance. In this chapter we focus our discussion on the evolution and interaction between the system of emission charges and permits used for protection of air quality, because they are among the oldest and most formalized policy instruments in the Hungarian environmental protection system.

The first general law on environmental protection in Hungary was adopted in 1976, but for years there had been a variety of laws and regulations whose purpose was to protect water and air quality as well as

---

1 Terminology related to environmental fines and charges can be confusing. In this chapter, we define an emission fine to be a charge per unit of emission that is assessed only upon exceeding of an emission limit and computed on the basis of the extent of the excess. This emphasizes the intent behind a fine: to punish exceeding of an emission limit. An emission charge, by contrast, may be assessed on the basis of any emissions, regardless of their relationship to an emission limit. All emission fine designs are emission charges, but only some emission charges are designed as emission fines and differences in design may mean great differences in the effect of these economic instruments in both principle and practice.

other media and natural resources. Subsequently several laws and regulations were passed or issued covering a variety of environmental media and problems. These laws were supplemented and strengthened in many respects after 1989, most especially by the passage of a decree in 1993 requiring environmental impact assessments (EIAs) and environmental permits for larger-scale activities affecting the environment. Together, these laws and regulations provided much of the basis for the system of fines and permits that we refer to as 'recent practice'.

In May 1995, the Hungarian parliament passed a broad piece of environmental legislation: Act LIII of 1995 on the General Rules of Environmental Protection (Act LIII, 1995). Commonly referred to as the new Environmental Framework Legislation (EFL), it superseded the 1976 EFL and took effect in December 1995. Legislation, decrees and regulations for implementing the new EFL have either only recently been issued or are still being developed and debated. Most indications are that a much more extensive and complex system of permits, standards, administrative requirements, charges and financial support for pollution protection will emerge from this process. Based on the language of the new EFL, draft legislation and discussions with knowledgeable individuals, this chapter also examines the prospects for change in the Hungarian environmental protection system, especially those affecting emission charges to protect air quality, and possible implications for the future operation and performance of emission charges as an instrument of environmental protection.

2 ORGANIZATION OF ENVIRONMENTAL PROTECTION

The Ministry of Environment and Regional Policy (MERP) has primary responsibility for environmental protection but a number of other ministries and organizations also have significant responsibility. A listing of organizations and their principle responsibilities under the new EFL, one which in many ways mirrors responsibility under current practice, is presented in Appendix 8A. As in most such systems, there are many areas of shared and overlapping responsibility.

A very important body of organizations with operational responsibility covering the environmental aspects of permits and fines are the Regional Environmental Inspectorates (Inspectorates). The Inspectorates perform a wide variety of administrative and official duties concerning air, water, hazardous waste management and noise protection, including monitoring and initiating actions against excessive emissions and levying and collecting associated fines. In Hungary there are twelve Inspectorates organized according to water catchment areas. The Inspectorates are the environmental protection organizations in direct contact with the individuals, firms and organizations that produce air and water pollutants, noise and hazardous wastes.

The authority of MERP and the Inspectorates derives from three sources: legislative acts (laws), government and ministerial decrees and regulations, and measurement standards. The principal sources of authority and recent or prospective revisions are:

1. Acts on:

   - environmental framework legislation (1976 superseded in 1995);
   - water management (1964 superseded in 1995);
   - nature conservation (1982 with prospective revision in 1996); and
   - product charges (1995).

2. Orders, decrees, regulations or decisions of the government or the different ministries on:

   - responsibilities of ministries (1990);
   - air quality protection (1973 superseded in 1986 with prospective revision in 1996);
   - protection against noise and vibration (1983 with prospective revision in 1996);
   - water management (1984 with prospective revision in 1996);
   - hazardous waste management (1981 with prospective revision in 1996); and
   - environmental impact assessments (1993, superseded in 1995).

3. Hungarian standards on sampling, measuring and analysing the different pollutants of air, water, soil, noise and vibration, and so on.

Detailed regulations on soil and subsurface water protection are being prepared but have not yet been issued.

3   STRUCTURE OF THE PERMITTING AND FINE SYSTEMS

While many elements of the general permit system in Hungary are functionally separate from the environmental fine system, permits can set technical, emission, monitoring and reporting conditions that complement, substitute or obviate the economic incentive or revenue functions of emission charges.

3.1 The Permit System

Environmental elements can arise as part of permitting in both the general system of construction and activity permits and the system of permits aimed specifically at environmental protection. The usual sequence of permits of environmental significance is shown in Figure 8.1.

There are many types of permits, but some of the most important, from an environmental perspective, are granted by the Technical Department of the local governments. Before the establishment of the EIA process, these permits, especially the construction permit, were the principal mechanism by which environmental protection considerations could be injected into the decisions of facility designers and developers. In reviewing these permits, the Technical Department is supposed to take into account national laws, regulations and policies as well as local conditions and concerns. Permits are required for:

- the use of land,
- construction of a building, and
- occupancy of a just-constructed building.

Inspectorates are designated as the 'expert organization' to review the environmental features of these permit applications and provide recommendations for conditions that should be met before granting permits. Provision

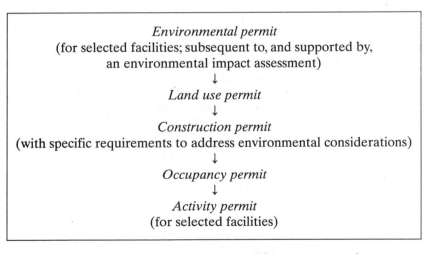

Figure 8.1   Permits with required or possible environmental protection content

specifically requiring consideration of environmental effects was added to the regulations governing issuance of construction permits in 1986.

There also exist other general permit requirements that include environmental elements. For example, regional water boards are required to issue permits before an organization can abstract, use or discharge water and the Ministry of Agriculture must issue a permit before a foodprocessing plant can be operated. In these and other cases of activity permitting, the Inspectorates have again been designed as the 'expert organization' for consideration of the environmental issues surrounding the granting of a permit.

The system of special environmental permits is based on the EIA regulations (1993, and modified in 1995). Under the regulations, qualifying new or modified facilities must provide a preliminary and, if necessary, a detailed EIA and obtain a special environmental permit from the appropriate regional Inspectorate before even requesting other general permits for use of land, construction, occupancy or production. Thus the EIA and its associated permit have constituted a necessary condition for many large-scale investment projects since 1993. Qualifying facilities under the EIA regulations include:

- large-scale industrial plants or production technologies as specified by activity and, in most cases, threshold production sizes in an Annex to the regulation,
- hazardous waste-processing facilities, and
- activities involving the incineration of wastes.

An important aspect of EIA-related permits, one that contrasts with the system of general building or occupancy permits, is that a new permit must be obtained if production or emissions increase by more than 25 per cent of the levels originally permitted.

As part of recent practice, approval of an environmental permit is based on a philosophy of prevention reinforced by the assumption that control incorporated into original designs provides cheaper environmental protection than add-on controls, remediation or averting behaviour. The permit applicant is required to present design or technology options and the Inspectorates select from among these options when establishing the conditions of the permit. The basis for selection is not codified, but when making their decision Inspectorates typically anchor on production or control technology in use or available in Hungary in combination with concern for meeting emission limits and ambient quality standards. The Inspectorates will only approve construction or other permits if they believe that the facility as planned will be able to meet its emission limits. In this sense, that is via facility design, emission limits are incorporated into the permit process.

3.2 Setting Emission Limits

Under environmental laws and regulations, Inspectorates must prescribe emission limits for common air pollutants and noise. Threshold limits for water pollutants are set by regulation. These emission limits become the basis for computing emission fines in Hungary. This system of emission limits and fines exceeds the permit system in scope. Emission limits and any associated fines apply to any emitting facility, including facilities built and operated many years before the advent of building, activity or environmental permits.

Permits and fines are interrelated in a number of ways. First, emission limits are frequently used as a reference point when Inspectorates review an EIA for an environmental permit or other permit applications. Facilities that have received environmental or other permits and emit air pollutants are also subject to higher fine levels for continued exceeding of emission limits. Finally, permits and fines together provide a combination of constraints and incentives and jointly help determine the level and cost of environmental protection at an emitting facility.

Emission limits need to be set by Inspectorates for approximately 330 air pollutants.[2] Of these, about 150 air pollutants have established measurement protocols that can be used to verify compliance and compute fines when emission limits are exceeded. When an existing facility substantially exceeds its emission limits, the Inspectorates can ask for an action plan from the operator and, based on this, can determine conditions that will ensure the facility can meet the emission limits in the future. These conditions include, for example, specification of operating conditions on control technologies such as precipitators; requiring installation of equipment, such as the central collection and isolation of polluted air streams; and management practices. If such conditions are determined to be practical, the Inspectorate can issue an 'authoritative obligation' for the facility to make the changes proposed. Except when there is an immediate threat to human health, the Inspectorate does not have the authority to impose additional penalties on the facility, but can impose an additional fine on a responsible individual if the remedies proposed by the Inspectorate are not met by the facility.[3] Issuing such an 'authoritative obligation' does, however, provide strong support for a

2   The limits described here apply to point sources of pollution. There are separate limit calculations for nonpoint sources, called building and diffuse sources.

3   In the case of an immediate health threat, the Inspectorate can close a facility for a while or restrict production. In less harmful cases these restrictions are not allowed under recent practice.

facility's application for grants or soft loans from the Central Environmental Protection Fund.

For air pollutants, regulations provide Inspectorates with formulas and tables to be used in the calculation of emission limits for various pollutants from different sources in different regions. This makes the calculation of emission limits for many sources fairly mechanical for Inspectorate staff; calculations have been computerized for more than ten years. The Inspectorate does, however, have the latitude to impose emission limits that are stricter than those calculated if ambient air quality limits will not normally be met using the calculated emission limits.

The procedure developed for actually determining the emission limit of a point-source air pollutant is summarized below. The emission limits, EL_{ikljn}, set for each point source n at a facility, are usually computed using equation (8.1):

$$EL_{ikljn} = K1_{ikl} * K2_{ik} * Ef_1/N_{lj}$$ (8.1)

where the limit itself is a function of:

- the air pollutant in question, i;
- the ambient levels of i in locality k in which the facility is located;
- the area protection priority assigned to area k;
- the height classification, l, of the stack or point source, n, that emits the air pollutant; and
- the number of point sources, N_{1l}, in height category l at the facility being permitted, j.

EL_{ikljn}, the emission limit, is given in kg per hour and is the maximum allowable average hourly mass flow of the emitted pollutant from the stack n in the height category l at site j. It is determined as the product of the three terms on the right-hand side of equation (8.1) and we will discuss each in turn.

$K1_{ikl}$, measured in $\mu g/m^3$, is the ambient air quality limit of the pollutant i based on the Ministry of Welfare's consideration of the health and environmental effects of particular pollutants and the degree of protection to be provided to various areas. For example, areas with major recreation and nature conservation sites are classified to receive higher levels of protection. These ambient limits are published as tables that vary with the pollutant and area classification. Table 8.1 illustrates these values for several prominent air pollutants and the three categories of area protection: Strictly Protected, Protected I and Protected II. The data in this table reflect the ambient air quality limits averaged over a 24-hour

Table 8.1   *Ambient air quality limits K1$_{ikl}$ by area classification for selected air pollutants*

Pollutant	Grade of hazard	Ambient air quality limit $\mu g/m^3$		
		Strictly protected area	Protected area I	Protected area II
		24-hour		
Sulphur dioxide	3	100	150	300
Carbon monoxide	2	2,000	5,000	10×10^3
Airborne dust	3	60	100	200
Soot	1	50	50	150
Nitrogen oxides (such as NO_2)	2	70	150	200
Nitrogen dioxide	2	70	85	150
Fluorides* a		5	5	5
b	2	10	10	10
c		30	30	30

Notes:
The grade of hazard is used to determine the fine (see Table 8.4).
*a. Fluorides in gas form, such as F (HF, SiF_4).
 b. Inorganic fluorides of airborne dust well soluble in water (NaF, Na_2SiF_2).
 c. Inorganic fluorides of airborne dust less soluble in water (AlF_3, $Na_3 AlF_6$, CaF_2).
Source: Extract from Hungarian Standard: Requirements of Cleanness of Ambient Air,
Table 1. MSZ 21854–1990 with Corrections.

period as set by the Ministry of Welfare. The Ministry also publishes
ambient limits based on annual and 30-minute time intervals, but the
24-hour limits are used to calculate EL_{ikljn} and, indirectly, the fines.

$K2_{ik}$ is a measure of the fraction of ambient air quality capacity 'available' in the region for use by the facility applying for a permit. It is a
unitless ratio computed with the formula $(100 - L_{ik})/100$, where L_{ik} is a
load index that varies with the category of the pollutant and the locality
in which the facility is located. Localities that already have higher ambient levels of a pollution receive higher pollution indices, resulting in
correspondingly lower values for $K2_{ik}$. Table 8.2 displays load indices for
the three categories of air pollutants, the three most common load index
combinations reflecting regional sensitivity, and the range of indices assigned for each pollutant category.

Table 8.2 *Load indices (L_{ik}) by pollutant group and common locality combinations for air pollutants*

	Load indices			
Type of substance	Most common load index combinations assigned to localities			Other load index combinations (variations within the range given)
Solid (particulates)	20	40	60	30–80
SO_2, NO_x, CO	10	30	50	20–70
Others	10	30	50	20–70

Note: Load indices vary for each settlement or by districts for Budapest.
Source: Based on 4/1986 (VI.2.) OKTH Order, Appendix 1, Table 1.

The numerator of the third term of equation (8.1), Ef_1 is a tabular value provided by MERP in the implementing regulations. Ef_1 links the emissions from stacks at height 1 to changes in ambient pollutant levels. Consequently, this factor has units of kg-m³/hr-µg. Dividing by N_{1j}, the number of stacks in that height category at site *j*, the emission factor limits are corrected for multiple sources at that height and site. The emission factor Ef_1 varies inversely and sharply with the height of the source. This is illustrated by Table 8.3, which contains the scale used to adjust emission limits for height.

Table 8.3 *Emission factors (E_{f1}) by source height for point-source air pollutants*

Emission height H (m)	Emission factor Ef_1 (kg/h*m³/µg)
$0 < H \leq 10$	0.002
$10 < H \leq 20$	0.006
$20 < H \leq 35$	0.09
$35 < H \leq 50$	0.70
$50 < H \leq 80$	2.0
$80 < H \leq 100$	4.0
$100 < H \leq 120$	6.0
$120 < H$	30.0

Source: Extract of 4/1986 (VI.2.) OKTH Order, Appendix 1.

3.3 Pollution Fine System

The pollution charge system in Hungary was designed as a fine. That is, it is only applied against those sources that exceeded their emissions limit as established by the Inspectorates. When an excess occurs, the fine is levied only on the difference between the actual emission and the calculated emission limit of the pollutant. Fines for exceeding an emission limit are applied to noise and vibration, approximately 150 air pollutants and 32 water pollutants where the method of measuring emissions is standardized.[4]

The fine in any quarter year, F_q, is based on average quarterly emissions, E_{ikljnq}, as reported annually by the operating organization for each source at each site. These reports are made to the Inspectorates of the region in which the site is located. The quarterly average emissions used in the report is based on adjusted hourly emissions for the actual hours of operation in quarter year q detailed in equation (8.2):

$$E_{ikljnq} = (\Sigma_h E_{ikljnh}) / T_q \qquad (8.2)$$

where E_{ikljnh} are the actual emissions, in kilograms for each hour of operation in the quarter year, and T_q is the number of hours of operation in that quarter year.

The actual fine is computed according to equation (8.3):

$$F_q = (E_{ikljnq} - EL_{ikljn}) * T_q * b_q \qquad (8.3)$$

where for each source, n, and pollutant, i, the fine is the product of the average rate by which the source exceeded its permitted emissions limit of i, EL_{ikljn}; the number of operating hours in the quarter, T_q; and a fine rate for the quarter, b_q, in HUF/kg. The structure of fine rates was set by MERP. The most recent rates for air pollutants are listed in Table 8.4 and have been in place since 1986.

---

4 Fines can also be assessed for violating certain rules or prohibitions stated in laws or orders related to environmental protection. Examples include:

- violating the rules on hazardous waste treatment (the fine depends on the quality and quantity of the waste, the treatment rule, and the duration of the offence);
- violation of prohibitions related to air quality protection, for example, neglecting the use of control equipment and burning wastes in the open or in conventional boilers not specifically constructed for waste combustion;
- failure to report the emission of air pollutants, the origin of hazardous waste, or make the yearly self-report on air pollution or hazardous waste treatment.

This fine rate b_q is a function of the degree to which the permitted emission limit is exceeded. As shown in Table 8.4, the fine rate rises in direct proportion to the degree of excess until the excess and fine rate are a factor of four times the initial levels. After that, the increase in b_q is less than proportional to excess. Note that under equation (8.3) if a step in the fine rate is triggered by an additional unit of average pollutant emissions, *the higher fine rate applies to all emissions in excess of the emission limit.*

The fine rates for air pollutants also vary with a number of other factors. For older facilities that were never required to apply, or applied before 1986, for a permit, the fine rates are increased by up to 80 per cent for repeated exceeding of emission limits over a period of four years. By contrast, the fine rates are escalated by two, three and four times when a facility with repeated excess received a permit since air quality regulations issued in 1986.

Table 8.4   Fine rates (b$_q$) for exceeding air pollutant emission limits

Factor by which actual emissions exceed the emission limit[3]	Fine rates according to the grade of hazard of the air pollutant[1] (HUF/kg)[2]			
	Grade 1 expressly hazardous	Grade 2 hazardous	Grade 3 moderately hazardous	Grade 4 practically not hazardous
1.00–2.00	1.0	0.5	0.3	0.2
2.01–4.00	2.0	1.0	0.6	0.4
4.01–8.00	4.0	2.0	1.2	0.8
8.01–12.00	6.0	3.0	1.8	1.2
12.01–20.00	8.0	4.0	2.4	1.6
20.01–50.00	10.0	5.0	3.0	2.0
50.01–100.00	12.0	6.0	3.6	2.4
100.01–	14.0	7.0	4.2	2.8

Notes:
1. These fine rates are multiplied by operational hours during the quarter to obtain the fine itself.
2. $1.00 = HUF 125.69 (1995 yearly average National Bank of Hungary exchange rate) calculated for each pollutant individually.
3. This ratio is the average hourly emission in a quarter year divided by the emission limit (both are in kg/h).

Source: Extract of 4/1986 (VI.2.) OKTH Order, Appendix 3.

The Inspectorate calculates the total fines based on annual emission self-reports. Facility fines are summed across all pollutants and sources at a site and an aggregate site fine is assessed against the firm operating the facility. Organizations can apply to the Inspectorate to receive a reduction in assessed fines based on amounts being invested in new emission control measures. So far, this is an option that has rarely been used.

3.4 Other Environmental Protection Instruments

Aside from permits and fines, the Inspectorates can also intervene in the operation of facilities by issuing 'authoritative obligations' that request pollution reductions, installation of control equipment, or require the development of an action plan identifying necessary measures, investments, and so on. This is usually done by an Inspectorate if a facility frequently exceeds its emission limits by a significant amount. This instrument is based on protection of public health, but the processes of facility selection and review and remedy determination are not formally established. Before passage of the 1993 EIA requirements, issuing an authoritative obligation was the primary means of environmental protection should the emission fine and permit systems fail in some respect. After 1993, they were used to address problems arising at facilities that were outside the EIA's scope, such as older facilities.

The required development of an action plan has, apparently, been a particularly favoured device for establishing an operator-determined schedule (with deadlines), financial commitments and personal responsibilities for improving environmental performance. The action plan is reviewed and accepted by the Inspectorate so that the environmental protection demands of the Inspectorate and the intentions of the operator can be coordinated, but compliance by permitted facilities is voluntary (see above).

4 ISSUES OF PERFORMANCE ARISING FROM RECENT PRACTICE

In this section we identify and discuss issues surrounding the performance of the mixed system of permits and fines just described. We wish to emphasize that many of these issues are not unique to Hungary; they are often an inherent part of any environmental protection policy. In this discussion our objective is to describe how, and to what extent, these issues manifest themselves in recent Hungarian practice.

We group issues according to environmental, financial and economic performance in order to help organize the discussion. In truth, these aspects of performance are related and often interdependent. We have also tried to provide quantitative and meaningful measures of the performance of recent practice. Unfortunately, data on many aspects of performance are difficult to obtain and their significance is often qualified by the data collection process and coincident changes in the structure and level of economic activity. Again, these difficulties are not unique to our examination of Hungary, but they should be borne in mind as the reader reflects on the issues raised and information provided in this section.

4.1 Environmental Protection

It is commonly argued that fine levels have been set too low in Hungary to effectively encourage polluters to reduce emissions to emission limits. While, from an economic perspective, failure on the part of some organizations to reduce emissions below emission limits does not necessarily mean that the fine levels are too low (see Section 5.2, below), it is unlikely that this has generally been the case in Hungary, at least for air emissions. Table 8.5 contains information on the 'base' level of fines in 1995 forints

Table 8.5   Selected pollutants and base fines for exceeding emission limits

	Base fine			
	HUF/ton (1995 HUF)		US$/ton (1995 $)	
Pollutant	1985	1995	1985	1995
Sulphur dioxide	857	300	6.82	2.39
Airborne dust	857	300	6.82	2.39
Nitrogen oxides	1,430	500	11.4	3.98
Ozone	2,860	1,000	22.7	7.96

Notes:
–The rates given are the base fine rates when the actual emissions exceed emission limits by a factor of 1.00–2.00.
–The 1985 rates are the rates expressed in 1 January 1995 HUF by using the industrial producer price index 1985–94. (The nominal rates remained unchanged since their introduction.)
–The 1985, 1995 US$/ton rates were both calculated with the yearly average US$ exchange rate of the National Bank of Hungary in 1995.

Source: Using the information in Hungarian Standard: Requirements of Cleanness of Ambient Air, Table 1, 3; OKTH Order 4/1986, Appendix 3; and Industrial Producer Price Index by Central Statistical Office; yearly average US$ exchange rate for 1995 from the Hungarian National Bank.

and dollars per ton. The fine levels are extraordinarily low by comparison with damage estimates commonly used for policy purposes in Western Europe and the United States. Kaderják's (1996) recent survey of such damage estimates made for air pollutants presents a broad range of values but estimates commonly exceeded $1,500/ton for SO_2, $2,400/ton for particulates (TSP), and $2,000/ton for NO_x. Even allowing for substantial downward adjustment in the damage estimates appropriate to Hungary's economic circumstances, the fine levels of recent practice are too low.

Even when fines are compounded by multiple pollutants and penalties that increase fine rates by up to a factor of four, it is unlikely that firms will find it cheaper to employ any but the most elementary of control technologies rather than pay the fines. By way of illustration, the coal-fired power station in Pécs pays a fine of 80 million HUF/year (approximately $600,000) for exceeding emission limits (primarily for SO_2). SO_2 control at the Pécs plant necessary to meet the emission limit is estimated to cost between ten and twenty billion forints ($75–150 million).

Whatever incentives fines can provide has been eroded over time because of inflation. The fines for air pollutants were established in 1986. From 1986 to 1994, the Industrial Producer Price Index rose by a factor of 2.80 and in the two years since then annual inflation rates were well above 20 per cent. Table 8.5 shows how, over a ten-year period, the real value of the base fine per metric ton has eroded for selected air pollutants. The inflationary environment recently experienced in Hungary (general price inflation in 1995 was nearly 30 per cent), in combination with the many possible legal motions and delays available to a polluter disputing a fine (see below), have over time sharply diminished the environmental protection afforded by the fine.

The fine system of recent practice has also provided low-cost options for avoiding fines that do not involve actually reducing pollution emissions. Consider, for example, the air emission fine described above: the value of Ef_1 for any given air pollutant in equation (8.1) is taken from a table that has the height of the stack as one dimension (see Table 8.3, above). The recommended emission limit increases rapidly with height and results in a much larger permitted emission from higher sources! This encourages the building of higher stacks which, while resulting in greater dilution of pollution, encourages larger emissions by weight and long-distance transportation of pollutants to receptors. As a result, larger sources, including fossil-fuelled power plants and smelters, have relatively high emission limits and pay relatively low fines.

Polluters have been able to delay, reduce or eliminate payment of fines by using the opportunities to appeal against a fine. In the inflationary environment described above, this can be an attractive option even if the chance for a successful appeal is low. At the same time, compromises, waivers and extensions are frequently granted when high fines accumulate or firms violating emission limits suffer financial setbacks. For example, in the late 1980s power plants sought and received a two-year exemption from progressive fine escalation for exceeding air emission limits.

Another means polluters have for exceeding emission limits without suffering the financial pain of fines is to declare bankruptcy. This seems to be a particularly attractive option for firms specializing in hazardous waste management. Such firms leave a legacy of both pollution and unpaid fines. Some financially weak enterprises do this by default, but others appear to adopt this tactic by design.

A system of fines, no matter how high the rate, cannot effectively reduce emissions to the emission limits so long as the organizations subject to fines can, in effect, finance the payment of fines from state budgets (Fucskó, 1993). This was particularly a problem before privatization of many of the industrial facilities in Hungary. It remains a problem for those state-owned or private enterprises that continue to get direct and indirect budget supplements from the state, or state-controlled organizations, to cover costs.

A potential problem with any system of charges as a pollution deterrent is the need to credibly and economically monitor and/or estimate actual emissions. Since pollutant emissions and operating times used to compute E_{ikljnq} are self-reported in Hungary, there is an incentive to underreport. Since emissions and operating levels are reported only once a year, the self-reports may be difficult to verify. Escalation of fines by up to 80 per cent with repeated excess increases the incentive for problem facilities to underreport in at least some periods. There is also no legal prescription to record any documentation related to the calculation of E_{ikljnq}, further reducing the incentive to accurately report emissions.

The current practice of Inspectorates regarding monitoring emissions or verifying self-reports has been to emphasize monitoring of larger facilities and facilities whose emissions are near or exceed emission limits. This capitalizes on one of the more tractable aspects of emission charges as a fine system: only sources that are in danger of exceeding emission limits need monitoring since the others have little or no incentive to underreport.[5] Emissions monitoring and inspecting are supported by

5   There may, however, be a strategic incentive to underreport emissions if there is a strong expectation that emission limits may be reduced or that parallel command-and-control requirements may be based on reported emissions. These are probably realistic concerns.

trained Inspectorate staff and modern equipment that appears to compare well with many Western European countries.

The protection of air quality provided in construction permitting has been modest because the opinion of the Inspectorates is usually solicited only during construction planning. Inspectorates are not generally involved in the issuing of occupancy permits and, even when they are, they frequently have only a sketchy idea of how the building and any installed equipment will actually be put to use and how day-to-day operations affecting air emissions will be managed. Only in the special cases where separate operating permits are also required (for example, for power stations, gas service and food industry plants, and for water use) can Inspectorates take into account the actual working conditions and environmental effects of the activity and make recommendations to the issuing ministry or state agency.

Recent moves to extend privatization of the Hungarian economy have stimulated changes in technologies, design, operation and production that can have important environmental implications. Recent practice in construction and activity permitting has been slow to deal with this problem; many previously permitted facilities either are not required, or contend they are not required, to request a new or revised permit when their operations change. Their permits were issued with the expectation that state-owned enterprises would operate for long periods with stable technology and production; little provision was made for adaptation and change that came with the transition to a market economy.

If a facility breaks the conditions of a construction, activity or environmental permit, it may be subject to fines but, so long as emissions stay within emission limits, the emission fines are not affected. These 'offence fines' (as opposed to 'emission fines') often reflect the tie between permits and regulations or ordinances. For example, offence fines can be levied against a facility for burning solid waste in its boiler when it was only permitted to burn conventional fuels or for bypassing a built-in control technology when the permit required its use.

Finally, although emission limits may be introduced into a permit, they have no more force to influence behaviour than do the emission fines in and of themselves. The only 'compounding' of penalties for violating emission limits tied to permitting is the higher penalty for repeated violation that applies to facilities that have a permit under the 1986 air pollution regulations.

The formal activity level of the Inspectorates under recent practice is summarized in Figure 8.2, which shows the decisions issued in the period 1991–93. This shows that the overall activity has risen steadily, but that most of the increase is due to actions pertaining to water protection.

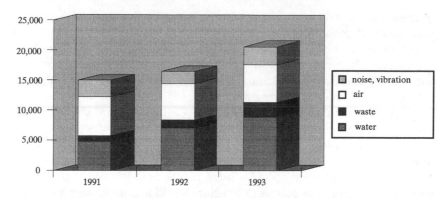

Figure 8.2   The number of decisions made by the Regional Environ-
mental Inspectorates by media: 1991–1993

Unfortunately, we do not have data indicating how many of these decisions reflect determination of emission limits, emission fines assessed,
permits or expert opinions issued, or authoritative obligations required.

4.2 Environmental Financing

Under recent practice, organizations that violate emission limits and pay
fines contribute to environmental protection if the revenues generated
from the fines and paid to the government are recycled into effective
environmental protection projects. The total fines imposed by media in
the period 1991–93 are shown in Figure 8.3.

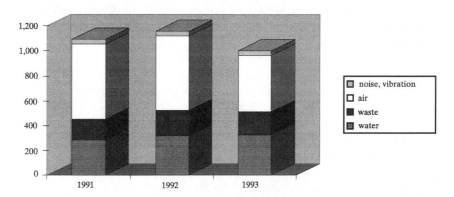

Figure 8.3   Fines imposed for different media by the Regional Environ-
mental Inspectorates: 1991–1993 (million HUF)

Recalling that the exchange rate in 1993 was approximately 90 forints to the dollar, these data do not indicate the availability of particularly large sums given the estimates of resources required for cleaning up past contamination and reducing current emissions.

Under the 1976 EFL, the revenue from fines was earmarked for environmental protection. The actual system of disbursement involves complicated formulas and earmarking based on both various categories of use and the origin of funds. A crude summary is that the fine revenues are commonly distributed 70:30 between the Central Environmental Protection Fund (CEPF) and the local governments in whose jurisdiction the fines originated. This distribution was determined in Hungary by a succession of budget laws. Table 8.6 presents ten years of data on the fine revenue distributed to the CEPF.

It is important to understand that these data reflect fine revenues in the year in which they were paid, not in either the year or amount in which they were assessed. These figures include all fines from violating environmental laws and limits, not just emission fines. Even so, we believe that most of these revenues derive from emission fines in the case of air and wastewater pollution and noise. The jump in revenues collected in 1988 is thought to reflect the new air regulations of 1986. Finally, these figures do not reflect assessed fines that were waived because of environmental investment activities undertaken by the operator subject to the fines. The spike in fine revenue in 1991 and lower levels of collection in subsequent years is thought to be due to (1) the transfer of fine administration to the Inspectorates at the end of 1990, (2) a two-year waiver of fine escalation granted to power plants, and (3) changes in economic structure and activity associated with the economic transition.

The resources of the Central Environmental Protection Fund are to be used to subsidize: control equipment, environmentally-friendly technologies and products, implementation of action plans or obligations based on authoritative obligations, processes reducing raw material and energy consumption, environmental protection research and development activity, education, training and technical development, and preventing and eliminating environmental damages. The conditions for applying for these funds as well as subsidy options available from the CEPF are specified in an order of the minister of MERP. Subsidy options include grants, loans with preferential rates, and loan guarantees. In the past decade, many investments have been assisted by the fund, for example, purchase and installation of control equipment, upgrading of waste-disposal sites, and transformation of vehicles to operate on natural gas.

Before 1990, the revenues from fines directed to municipalities accrued to Local Environmental Protection Funds and were earmarked for

Table 8.6 *Environmental fine revenue by media for the Central Environmental Protection Fund 1985–1995 (in million 1995 HUF)*

	1986	1987	1988	1989	1990	1991	1992	1993	1994	1995 estimated*
Air pollution	360	437	992	638	770	1,179	321	439	470	350
Hazardous waste	232	497	186	168	25	89	47	49	29	40
Noise and vibration	78	123	100	59	20	47	22	28	11	20
Nature protection	3	3	1	1	0	1	0	0	1	0
Wastewater**	–	–	–	–	–	236	291	174	152	200
Total	673	1,060	1,279	866	815	1,552	681	690	663	610

Notes:
–This table is calculated by using the industrial producer price index, calculated by the Central Statistical Office.
–US$1 = HUF 125.69 (1995 yearly average National Bank Hungary exchange rate).

Sources: 1986–1994: Ministry of Environment and Regional Policy, Department of Economics, 1995.
* Source 1995: *State Budget of Hungary* (1996) forecast.
**Before 1991 fine revenues went to the Water Fund. After 1991 the Water Fund got revenue from the water abstraction charge.

environmental protection, for example, renewing polluting boilers and public transport buses, investing in ambient air monitoring and central heating system development. Since 1991, local municipalities' use of these funds has been unrestricted. Given the fiscal difficulties occasioned by economic transition, most local community governments have chosen to use most of these funds for purposes other than environmental protection.

4.3 Economic Performance

The first two parts of this section examined recent practice from the perspective of emissions reduction: recent practice as a vehicle for reducing emissions and financing those emission-reducing activities. Here we briefly examine the economic issues measured by the efficiency and cost effectiveness of environmental protection.

The Hungarian fine system might superficially appear to resemble what neoclassical economists would call a Pigouvian tax or environmental charge – the typical reference point for economists when examining economically efficient internalization of an environmental externality. For a variety of reasons, however, it fails to meet the requirements of a Pigouvian tax, and hence does not create the economic efficiency-related advantages of such a charge.

An important difficulty, especially in the era before the 'change in system', is the distortion in prices caused by a centralized system of resource allocation. In such an economic environment, prices can be misleading indicators of social value and one does not know whether they overvalue or undervalue the 'marketable' resources embodied in them. Consequently, one cannot know what correction, if any, should be applied for 'non-market resources' (such as the environment). In such circumstances, a fine system was simply a means of enforcing emission limits; a means whose economic merit is impossible to assess in a straightforward manner because of the difficulty in knowing the economic value of resources used to protect the environment and commodities produced jointly with pollution.

With the transition to a market economy, the possible role of the fine as a 'Pigouvian' tax becomes more meaningful. Unfortunately, the current fine system does not possess other elements of an efficient Pigouvian design. First, while there are some efforts to vary charge rates with damage, for example, the sensitivity of the locality and the toxicity of the pollutant, the charge levels were set without anchoring on valuation of marginal damages. The economic efficiency of a Pigouvian tax depends critically on linking a tax or charge rate to damage estimates. An emissions fine that is not based on such damages is likely to be economically inefficient. Second, fines are a step function of source emissions; no fines are paid unless the emission limits are exceeded and then the fine is paid on each unit of emissions above the limit. This means that both within a source and across sources different fines are being paid for producing the same pollutant. Thus, assuming that every unit of pollution causes the same damage, the fine system, as employed in recent practice, is not cost effective. Third, the penalty features of fines create discontinuities in the marginal fine structure, for example, the additional cost of a one unit increase in emissions over an emission limit may not be just the fine rate, but the fine rate plus the penalty on emissions above the emission limit in previous years. This, like the problem of the step function nature of the fine just discussed, contradicts the design principle of the Pigouvian tax and probably results in further economic inefficiency and loss of cost effectiveness. If we want to improve the economic performance, the design of the emission fines will have to be changed to eliminate these three shortcomings in economic design.

While in Section 4.1 we examined the merits and issues associated with using emission fines and permits to reduce emissions, it is possible that emission reductions can exceed that which is best from an economic perspective. For example, if emission limits are set too high, in the sense that meeting these limits requires costs in excess of the benefits (damages avoided), permits or emission fines that effectively result in organizations making the investments needed to meet the limits are economically inefficient. In this case, greater environmental effectiveness may result in less economic effectiveness. Conversely, it is possible that failures to reduce emissions effectively may have actually resulted in outcomes that, given economic conditions, are advantageous from an efficiency perspective. We do not contend that this was the case; in fact we think that it was unlikely, especially in the period preceding the 'change in system'. We simply want to emphasize the importance, when assessing the performance of an environmental protection system, of balancing the undoubted benefits of protection with the cost of obtaining those benefits.

By way of example, the provision of the system that allows stack height to enter into calculations of emission limits may result in greater emissions (a disadvantage of recent practice under environmental performance measures), but provide a better reflection of expected health and other damages associated with emissions released at different heights. Thus, when properly established, allowing emission limits to vary with stack height may actually be economically advantageous. Whether it actually is or not depends on the relationship between stack height, emission fines and economic damages in Hungary. Once more the point with respect to recent practice is that a problem from an emissions reduction perspective may be a boon from an economics perspective. We want to emphasize that this result is not due to *choosing* economic considerations *over* environmental considerations, but to a *balancing among* environmental and other considerations (such as consumption and goods production) that is at the heart of economic performance measures.

From an empirical perspective it is likely, as noted above, that fines are too low in the sense that environmental damages are typically greater than effective fine rates. In the absence of other policy instruments this would result in emission levels that are too high. In a mixed system that characterizes recent practice, especially one in which the permits and authoritative obligations can be brought to bear on polluters, we do not know if emission levels have been 'too high' without additional study. We can say, however, that it is quite unlikely, given the design of emission charges as fines, in combination with the exercise of permitting and authoritative obligations powers, that emission reductions were achieved in a cost-effective manner.

5 CURRENT PROSPECTS FOR CHANGE

5.1 Environmental Permits, Emission Limits and Ambient Air Quality

The 1995 EFL extends the use of EIAs and environmental audits as instruments of environmental protection. It also calls for the establishment of various technical, emission or ambient standards motivated almost exclusively by consideration of environmental protection and the technical possibilities for such protection. Much of the inspiration for the scope and design of these standards comes from European Union guidelines on environmental protection. These EIA and audit processes are likely to be extended under new, proposed or future regulations to a much broader range of facilities, including existing facilities that already hold general construction and activity permits but which were not formerly subject to the EIA process.

In thinking about emission limits, environmental authorities hope to close some of the 'loopholes' that characterized the past system. For example, air quality emission limits at sources will likely be set on the basis of concentrations without reference to the height of sources. These will be translated into kg/hr emission limits for the purpose of establishing emission charge schedules, but the height of the source will no longer be a basis for scaling up emission limits. Existing facilities will have a grace period of years after which time they must be able to meet technology-based emission limits as a condition of operation. The requirements will be given strength by allowing Inspectorates to force production cuts or closure if the emission limits are not met.

The 1995 EFL does not explicitly encourage balancing costs and benefits, or even considering costs, when setting limits or standards. Furthermore, though the new EFL does mandate development of an environmental information system and analyses of a wide variety of environmental impacts as part of an EIA, an environmental audit, or a law or regulations, it does not recognize formal consideration of economic or cost impacts as either necessary or desirable. While such balancing will inevitably become an implicit part of the process during required consultations with affected and interested parties, there is no suggestion that a professional analysis of the economic issues should be conducted and made public.

The impact of the new EFL on the actual performance of the system of permits and fines for environmental protection will depend on the implementing regulations. While, as noted above, most of these regulations are still being drafted, some consensus appears to have been reached on design philosophy for some of these regulations. For the new air regulations, there appears to be general agreement that permit conditions, and emission limits should depend on ambient air quality levels and the cost and availability of control technology. Bracketing this view is a dual notion that 'poor' or obsolete control technologies should not be permitted even if an emission limit would not be violated, and that the most sophisticated and expensive control technologies should be required only when ambient standards are jeopardized.

5.2 Charges and Fines

While subject to considerable interpretation, the 1995 EFL identifies four charges providing 'cover', that is, revenues, to support 'measures abating the loading and the utilization of the environment'. These charges are: environmental emission charges, user charges, product charges and deposits.[6] The criteria for designing these charges are:

- magnitudes . . . that encourage reductions in emissions,
- magnitudes and 'goals [the charges] are used for' to be set in 'conciliation' with those paying the charges,

6   The translation of these terms into English is not very conventional. The emission charge is translated as 'environmental load charge' and the user charge as 'utilization contribution'.

- 'goals and manner of using the charges' shall provide that the greater part of the revenue be spent on abatement of emissions that made up the basis for the charges paid.

The 1995 EFL also retains emission fines as a policy tool (Section 106). Those who violate regulations, decisions or standards *shall* pay a fine in 'conformity with the level, weight, and recurrence' of excess emissions and 'environmental damage caused'. This last phrase does, at least in the case of fines, link their level to the amount of damages, a feature missing in the EFL discussion of charges *per se*. Continued use of fines, on top of emission charges, however, creates possible redundancy and adverse economic and cost implications. The combination of punitive emission charges and fines for exceeding practically guarantees overcontrol of emissions, overpayment for the control that is obtained (due to failure to be cost effective across sources), and diversion of charge revenues from the most beneficial projects because of earmarking.

'Environmental emissions charges' are described as charges on all emissions, not just a charge on emissions exceeding emission limits (EFL, 1995, Section 60.4). From an economic perspective, this aspect of the emission charge may represent an improvement in design over the old fine system if the level of the charge itself and other design features of the charge are set properly. In particular, the emission charge described by the new EFL encourages organizations with low costs of control to continue reducing emissions beyond these emission limits. This is a particularly valuable property if, after setting charges and emission limits, there is a lot of variation in the control options both across and within organizations subject to the charge. In such cases, the described charge system can achieve a target level of environmental protection at a substantially lower cost than can the system of recent practice. Unfortunately, one gets the sense, given the other features of charge design and the criteria listed above, that this feature is intended primarily as a revenue device, that is, to provide revenues to the government even when a facility is meeting its emission limits.

Like the emission fine of recent practice, the EFL limits these charges to pollutants for which there are measurement standards and provides the legal basis for record keeping and self-reporting requirements. The legislation also says that emission charges '*shall* be specified separately' for different types of emissions and set proportionately to emissions and *may* vary with the 'category of the area and the emission standards'. Thus there is the possibility that emission charges will also be made a function of the emission limits with, as noted above, the attendant difficulties that such interpretation causes for the achievement of cost effectiveness in particular and economic efficiency in general. As we understand it, this is not the interpretation embodied in draft air regulations; there are expected to be no 'steps' in the emission charge schedule, but there may be different emission charge rates for different sources and pollutants depending on the circumstances. The economic merit of such distinctions will depend on the types of circumstances selected to trigger different rates.

Another common interpretation of the new EFL's mandate on emission charges is that it directs the government to set emission charges so high that even a facility with the highest costs of control will be compelled to make the investments necessary to meet emission limits. As long as the technology-based emission limits are feasible to obtain, a single emission charge applied against all sources posing comparable environmental or health threats will result in a cost-effective system of emission charges. While this may be true, it is unlikely to result in environmental protection that is economically efficient. By anchoring charge rates on facilities with the highest control costs, facilities as a group are highly likely to overcontrol relative to economically and socially desirable levels. Even if the revenue for the charge could be 'recycled' to pay for investments in control equipment, this approach may lead to a 'double loss': affected facilities reduce emissions too much and potential tax revenue is diverted to public investments with a low social rate of return. It is economically preferable to anchor the charge rate on the best estimate of the environmental damage avoided when emissions are reduced. There is no suggestion in the EFL that MERP should be guided by this principle when it 'conciliates' with those obliged to pay the charge.

The 1995 EFL directs that emission charges should be phased in, but it does not indicate how this phasing should be coordinated with the establishment or effective date of new or revised emission limits. The new air quality regulations are likely to give the local Inspectorates considerable latitude in the phasing-in of the new system of charges.

The new EFL does not provide procedures or guidance for escalation or adjustment of emission charges due to either erosion in their real value (due to inflation) or changes in estimates of the damages done by emissions. This will impair the system of charges over time if the charge structures and rates are optimally designed to begin with. The problem of erosion of the value of fines due to inflation has been partially addressed in the case of new emission charges by requiring polluting organizations to deposit the amount of the charge with the Chief Inspectorate if an appeal is made to the court. These funds can be deposited in a bank and receive interest. In this way, the future value of the charge will not be eroded, while waiting for a final judgment.

5.3 Environmental Financing

As noted above, the language of the 1995 EFL generally, and the section
on charges in particular, seems to emphasize the philosophy that charges
are revenue instruments (taxes) whose yield is to be earmarked to subsidize or 'cover' environmental protection projects selected by the central
or local governments. The various charges and fines identified in the new
EFL, taken together, could provide large multiples of the historic fine
revenues cited above, because the charges would be applied to all emissions not just to those exceeding emission limits. The actual impact on
revenues will, of course, depend in part on the charge rates set, but one
common school of thought is to set charge and fine rates that are much
higher than at present. The revenue impacts also depend on the effectiveness of enforcement and the emission reductions that actually occur. The
question remains, however, as to whether this makes fiscal or economic
sense.

The 1995 EFL requires no fiscal evaluation of revenue generation
using environmental charges or fines. This is a problem because it might
turn out to be more efficient to raise these revenues through direct
taxation (administrative costs for some of these charges and fines may be
very high). A fiscal evaluation may also help to understand the possible
economic toll of a strategy of high charge rates, even when coupled with
recycling of the revenues. Morris and Kis (1997) estimate the excess
burden of product charges on tyres to be 7–10 per cent of revenues.

The 1995 EFL requires that revenues from charges and fines be spent
on environmental protection projects and that the goals of these projects
be determined through 'conciliation' with those who paid the charges.
Indeed, as noted above, two of the three criteria that are meant to structure the environmental charges relate to disbursement. Additionally,
Sections 59(4) and (5) say that the 'goals and manner of using' charges
are to be determined by specific regulations, that the 'greater part' of the
revenues should be spent on control of pollutants that were the basis for
the revenues, and that the charges should be paid to the funds specified
by law. The use of charge revenues is, if anything, likely to be even more
narrowly earmarked in the future than under earlier legislation and
regulations.

This is certainly the case with the municipal share of charge and fine
revenues. The new EFL allows a portion of emission and user charges to
be distributed to local governments only when the local governments
create a separate, local Environmental Protection Fund whose resources
can only be used for the purpose of environmental protection. Such
earmarking assumes that local governments either cannot or will not

correctly determine the proper role of environmental protection in local spending priorities. This may be the case, but most public finance theory assumes that, in a democratic society, local governments are actually in the best position to reflect the group preferences for local goods and services. Perhaps most importantly, the disbursement guidance provided in the EFL does not require or even suggest that economic efficiency or even cost-effectiveness criteria be used in the award of subsidies supported by charge revenues.

6 CONCLUSION

Environmental charges in the form of fines tied to exceeding of emission limits have for some time been an integral part of the mixed system of permits, standards and operating requirements used in Hungary for the protection of environmental services. These fines have not, however, generally been effective as a means of ensuring that emission limits are met. Failure under recent practice is attributed to the low initial level of fines relative to control costs, the declining real value of the fines in an inflationary setting, and the opportunities for low-cost means of avoiding or reducing fines.

Environmental considerations have, over time, been added to the general construction and activity permit systems. These systems, however, have often been overwhelmed by the size and pace of change in the organizational structure of economic activity occurring over the last six years. A new class of specifically environmental permits has been developing tied to the environmental impact assessment process. It is too early to judge the performance of this process and the resulting environmental permit system, since the EIA process only began in 1993.

Recently the Hungarian parliament and the ministries have been attempting to remedy the problem of insufficient environmental protection with the passage of new laws and regulations. While many of these initiatives are still being developed or debated, the 1995 EFL provides the structure and impetus for much of this change. The new EFL and associated laws do, on paper, remedy or address some of the tactical problems identified with recent environmental protection practice. For example, a system of emission charges is supported and the Inspectorates can now require a deposit when a fine is appealed.

Strategically, however, the new EFL mandates potentially costly and cumbersome EIA and audit processes in combination with the development of new emission limits that are widely interpreted as principally based on either technical feasibility or best-industry practice. Just as importantly, the new EFL does not suggest or even allude to the merit of balancing economic costs and benefits when making environmental policy decisions. Formal consideration of this sort might substantially improve performance of any of the environmental charges the new EFL sanctions or the disbursement of revenues that these charges and fines generate.

REFERENCES

4/1986 (VI.2.) OKTH Order: Regulation about the Execution of 21/1986 (VI.2.) Council of Ministers' Order on Clean Air Protection.
86/1993 (VI.4.) Order of the Hungarian Government on Transitional Regulation of Environmental Impact Assessment of Certain Activities.
152/1995 (XII.12.) Order of the Hungarian Government on the Scope of Activities Bound to the Execution of Environmental Impact Assessment and the Detailed Regulations of the Related Procedures of the Authorities.
Act LIII (1995) on the General Rules of Environmental Protection (September 1995), English translation in 'Hungarian Rules of Law in Force' No. VI/17, 1995, Budapest.
Fucskó, József (1993), 'The Environmental Impact Assessment Process in Hungary', unpublished paper. Hungarian Standard: Requirements of Cleanliness of Ambient Air, MSZ 21854–1990 with Corrections (1990, corrections 1990–1995).
Kaderják, Péter (1996), 'Estimating the External Damages Caused by Air Pollution of the Power Sector in Hungary' (draft), Harvard Institute for International Development, Hungary Environmental Economics and Policy Program Working Paper, May.
Ministry of Environment and Regional Policy, Department of Economics (1995).
Morris, Glenn E. and András Kis (1997), 'The Design and Assessment of an Environmental Product Charge on Tires in Hungary' (draft), March.
National Bank of Hungary, Department of Economics (1996).
State of Budget of Hungary 1996 (1996).
Statistical Yearbook of Hungary 1994 (1995), Central Statistical Office, Budapest.

APPENDIX 8A   SELECTED ENVIRONMENTAL PROTECTION RESPONSIBILITIES IN HUNGARY

Responsibilities and jurisdictions of different organizational levels:

Ministry of Environment and Regional Policy (MERP)

1. direct:

    -   environmental protection activities,
    -   the implementation of international treaties, and
    -   environmental protection activities within MERP jurisdiction are the media or environmental resource: air, water quality and soil, protection from the harmful effects of hazardous wastes and substances, protection from noise and vibration;

2. analyse and evaluate:

    -   the state of the environment and the progress made in its protection,
    -   the process of natural resource management,
    -   experiences with protection activities, and
    -   activities to prevent and surmount environmental dangers and damages;

3. make a draft of the National Environmental Protection Programme;
4. participate in the development of regulations in connection with environmental protection (laws, governmental, and ministerial orders and so on);
5. operate the National Environmental Protection Information System (based on measurement, observation, and monitoring network to collect, process and record data);
6. coordinate the research and technical developments and cooperate in education and training related to environmental protection;
7. assist in the regulation of the economic bases of environmental protection (tax, customs, duty allowances, environmental charges, fines, and so on).

Ministry of Welfare

1. determine:

    -   ambient limits for air and water quality and develop permissible levels of noise and vibration, and
    -   classification of chemicals and hazardous wastes;

2. measure the ambient air, drinking, bathing and wastewater quality (executed by the county State Public Health Services).

Ministry of Transport, Communication and Water Management

1. manage water resources: quantity of surface and subsurface waters, use of waters and prevention of floods;
2. control the emission of air pollutants and noise of different vehicles (executed by the county Traffic Authorities and the Police).

Ministry of Agriculture

- control the use of the soil, pesticides and forests.

Environmental Inspectorates

1. coordinate the regional environmental activities performed by other organizations;
2. provide professional assistance and technical suggestions to the local governments, co-authorities, companies and so on;
3. collect and evaluate data measured by the Inspectorates or other laboratories and assure public availability;
4. make official estimates and forecasts of possible changes in the regional environment;
5. direct the necessary activity to localize and eliminate accidental contamination (water, soil, hazardous waste);
6. issue environmental protection permits based on EIAs;
7. give permits for the treatment of hazardous wastes (to collect, transport, deposit, convert, eliminate, incinerate and so on) and sewage;
8. provide expert, authoritative opinions during the proceedings of other government authorities (in land use, building, occupancy, and in special cases to put into operation a plant, and so on);
9. determine and prescribe the emission limits of air pollutants and for noise;
10. require emission reductions, control equipment or reconstruction of polluting technologies;
11. implement sanctions:

    - impose fines for exceeding permitted emission limits for water and air pollutants, noise or vibration,
    - impose fines for violating the rules on hazardous waste treatment or for neglecting any prohibitions stated in laws or orders (for

example, burning stubble-field or wastes in the open air, shut-down of control equipment),

- impose offense fines for neglecting notifications or failure to provide reports, and
- impose fines for accidental pollution of the air or water.

Local Community Governments

1. determine the zoning of different protected areas for air quality and noise;
2. implement the authoritative jurisdictions to service companies and entrepreneurs;
3. issue land-use, construction, occupancy and activity permits.

9. Environmental Charges in Poland

Glen D. Anderson and Boguslaw Fiedor

1 INTRODUCTION

Since the 1970s, Poland has utilized a system of environmental and resource charges or fines as part of its environmental management system.
However, only in the last five years have charges and fines made more
than a nominal contribution to the promotion of environmental goals.
The primary policy instrument for achieving environmental objectives is
the national system of facility permits for point sources of pollution
emissions and discharges. All point sources of pollution are required to
apply for and maintain pollution permits. These permits specify the allowable discharges of every regulated pollutant from every pipe, stack or
other type of conveyance. In principle, local environmental authorities set
allowable emission and discharge levels for each source in order to protect ambient air and water quality levels.

Historically, both charges and fines have been closely linked to the
facility permit. In the past, charges were levied on the allowable emission or discharge level specified in the facility permit instead of actual
levels. The charge was determined by multiplying the allowable emission
level or concentration for wastewater by the appropriate charge rate.
In the last few years, charges have been based on actual rather than
allowable emission or discharge levels. The change in approach can be
attributed to the dramatic increase in charge rates, wherein polluters
whose pollution levels are below allowable levels have a clear incentive
to base their charges on actual levels. Specific charge rates have been
developed for a variety of air pollutants, solid and hazardous wastes,
discharges of wastewater, including saline mining waters, surface- and
groundwater extraction, and land clearing. Fines can be issued if the
regulated firm discharges or emits pollution in excess of the levels specified in the facility permit. Generally, the fine is equal to the difference
between actual pollutant levels and permitted levels multiplied by the

fine rate, which is simply a multiple of the charge rate (up to ten times for air pollutants).

Until the end of the communist regime, charge rates were quite low and did not have a discernible effect on pollution levels or generate much revenue. Since 1990, the charge rates have been increased dramatically to approximately 18–20 times their levels during the communist regime and now are among the highest anywhere in the world. Revenues from charges and fines are distributed to the National Fund for Environmental Protection and Water Management, 49 regional (Voivod) environmental funds, and approximately 2,400 local (Gmina) funds. These funds, in turn, disburse charge and fine revenues for environmental investments and other environmental activities (for example, education, research, monitoring equipment) through a variety of disbursement mechanisms ('soft' or subsidized loans, grants and interest subsidies on commercial loans). Approximately US$450 to 500 million are collected annually from environmental charges and fines, enabling environmental funds at all levels to account for nearly one-half of annual capital costs of environmental investments in Poland.

Thus, the system of environmental charges and fines is an integral and important component of environmental management in Poland that complements direct regulation through permitting, provides modest incentives for reducing pollution, and makes a major contribution to the financing of environmental improvements. Of less importance than charges, fines and direct regulation in Poland's environmental policy, are other economic instruments such as differentiated taxes and duties on imported goods. These taxes often include provisions for lower rates for more environmentally-friendly goods. For example, the excise tax on unleaded gasoline is lower than for leaded gasoline, resulting in lower retail prices, even though production costs are higher for unleaded gasoline. Also, a significantly lower value-added tax (VAT) of 7 per cent is assessed for some environmental protection items compared to the usual VAT rate of 22 per cent. Finally, there is an assortment of minor investment tax credits and tax deductions for charitable contributions of an environmental nature.

The purpose of this chapter is to describe the system of charges, fines and permits in Poland, analyse the implementation experience to date, and identify potential or needed improvements. The next section provides a more detailed description of the respective charge, fine and permit systems, and highlights key features of each.

Section 3 focuses on practical aspects of implementation, while Section 4 provides an evaluation of these environmental systems. In the concluding section, some recommendations for improving the system of charges, fines and permitting are presented.

2 THE STRUCTURE OF THE SYSTEM OF ENVIRONMENTAL CHARGES AND PERMITS

2.1 Environmental Charges

The provisions for imposing environmental charges are contained in the Environmental Protection Law (EPL) of 1980, the Water Law of 1974, and amendments to each of these laws. Both acts include declarations to the effect that the economic use or modification of the natural environment provides the basic motivation for assessing charges. Thus, both the *user pays principle* and the *polluter pays principle* would seem to provide a rationale for charges. The EPL and the Water Law further restrict the use of revenues from charges to ecological purposes and require disbursement through environmental funds.

A broader statement of the role and purpose of environmental charges is outlined in the National Environmental Policy of Poland, approved by the Council of Ministers in 1990 and the Polish parliament in 1991. It is the stated intent of the government that charges should serve a complementary role in the national administrative/regulatory system of environmental management. The first objective of charges is to encourage polluters to minimize the social costs of meeting environmental goals. Thus, to meet this objective, charge rates must be high enough to encourage polluters to invest in control rather than to simply pay the charge. In addition, the charges should bear some relation to the marginal damages resulting from pollution emissions or discharges. The third objective of charges is to generate revenues which can then be recirculated for environmental investments and related purposes (for example, education, research, monitoring equipment). Environmental charges also are viewed as an integral part of Poland's commitment to the polluter pays principle and are expected to contribute to the fulfilment of Poland's short-, medium-, and long-term environmental priorities enumerated in the National Environmental Policy.

As noted earlier, charges are assessed for a variety of uses of the environment and natural resources. The procedure for promulgating environmental charges begins in the Ministry of Environmental Protection, Natural Resources and Forestry (MEPNR&F), where staff prepare the list of pollutants for which charges are to be assessed and develop proposed charge rates. The proposed charges and rates are submitted to the Council of Ministers for approval. Generally, charge rates are revised annually. Since 1992, the year when charge rates were raised to their highest real rates, annual revisions to the charge rates have been motivated by the desire to maintain the real – as opposed to nominal – rate structure of the charges, although there have been some minor revisions in specific charges as well. Slippage in charge rates due to inflation is largely avoided by adjusting nominal charge rates to account for projected inflation in the next year. If actual inflation deviates significantly in either direction from projected inflation, adjustments can be made in the following year.

For all charges assessed on emissions or discharges, the Ministry of Environmental Protection has attempted to group pollutants according to potential damage to human health and the environment. Generally, *relative* differences in rates are correlated with differences in pollutants' toxicities or potential to cause environmental damage. For example, solid and hazardous wastes are grouped into four categories according to the toxicity of the wastes. In addition, air emissions of heavy metals and suspected carcinogenic pollutants such as benzene are assigned extremely high rates. The *absolute* magnitudes of the rates are not set to reflect marginal damages to health and the environment or to correspond with the marginal costs of abatement. Such factors as damages, abatement costs and economic characteristics of the polluting sector are taken into consideration, but ultimately the rates are set at levels that are politically acceptable and meet revenue requirements.

There are two other types of charge-rate differentiation. First, charges for surface- and groundwater extraction are differentiated by geographical region. Poland has many regions with chronic water supply shortages. Thus, these charges vary according to the scarcity of water supplies. Charge rates also vary according to the type of use (for surface- and groundwater extraction and wastewater discharges). For example, the wastewater charge for biological oxygen demand (BOD_5) is ten times higher for power generation than for municipal sewage.

The types of environmental charges and the amount charged are described in the appendix, which is based on Directives 637 and 772 of the Council of Ministers, issued 27 December 1993 and 28 December 1994, respectively. In brief, air pollution charges have been developed for 62 specific air pollutants and similar or related compounds plus seven types of evaporative air emissions. For the eight most toxic air pollutants (acrylonitrile, arsenic, asbestos, benzene, benzo-a-pyrene, polychlorovinyl, chromium and nickel), environmental charges are PLN (new Polish zlotys) 123.6 per kilogram or about US$53,739 per ton. Charges for common air pollutants such as SO_2 and NO_x are US$82 per ton (see Table 9A.1 in the appendix). Charges are also levied on 163 types of solid wastes divided into four categories on the basis of toxicity. The charge for the four categories from most to least toxic wastes are US$21.49 per ton, US$8.06 per ton, US$2.69 per ton, and US$1.61 per ton, respectively (see

Table 9A.2 in the appendix). Charges for wastewater are related to concentration levels for BOD_5 chemical oxygen demand, suspended solids, chlorate and sulphate ions, heavy metals and volatile phenols (see Table 9A.3 in the appendix).

Environmental charges normally are collected once a year by the regional (Voivod) environmental protection departments. As a result of a 1990 amendment to the EPL, the Governor of the Voivod may request that large enterprises pay their charges in quarterly installments. Enterprises are able to treat environmental charges as normal business expenses and to deduct the amount of charges paid from taxable income. Thus, charges are treated as a normal production cost. An interesting provision allows enterprises to deduct the amount of the charges *levied* in the current year from the current year taxable income even if the enterprise is delinquent in making payments and pays the charges in the next calendar year.

Once charges (and noncompliance penalties) are collected, they are distributed to the National, Voivod and Gmina environmental funds. Three alternative distribution schemes are used to allocate revenues to the three types of funds. Most charges and fines for air pollution, water extractions and wastewater discharges are distributed according to the first distribution scheme (Table 9.1). These air and water charges and

Table 9.1 Distribution of environmental charges and fines

Three distribution schemes

1. Charges and noncompliance fines on air emissions (except NO_x); charges for surface- and groundwater extractions; wastewater charges and noncompliance fines:

National Environmental Fund	36%
Voivodship Environmental Funds	54%
Gmina Environmental Funds	10%

2. Charges and noncompliance fines on NO_x; charges and noncompliance fines on saline-mining wastewater discharges:

National Environmental Fund	90%
Gmina Environmental Funds	10%

3. Charges and noncompliance fines on solid and hazardous waste disposal:

National Environmental Fund	20%
Voivodship Environmental Funds	30%
Gmina Environmental Funds	50%

Source: Spyrka (1994).

fines represent the largest share of revenues. As a result, the overall shares of total revenues distributed to the National, Voivod and Gmina funds is almost identical to the share allocations for the first distribution scheme. The bulk of charges and fines for nitrogen oxides and saline-mining wastewater discharges are distributed to the National Fund, while the greatest share of hazardous and solid waste charges and fines are distributed to local and Voivod funds. The impetus for distributing a large proportion of the revenues for nitrogen oxides to the National Fund relates to the desire to encourage cost-effective improvements through-out the country (since nitrogen oxides are a regional and transboundary pollutant). Most revenues from saline-mining water charges and fines go to the National Fund to provide flexibility for financing investments in controls at mines, as well as treatment in downstream river basins. Similarly, the distribution scheme for solid and hazardous waste charges and fines reflects the local nature of disposal and impacts.

All charges and fines distributed to the National and Voivod funds must be *earmarked* for investments and other expenditures to reduce the general types of pollution from which the revenues were collected. Thus, wastewater charges and fines (excepting saline-mining waters) must be dedicated to reducing wastewater discharges, but not necessarily to address a specific type of wastewater problem (such as suspended solids). Only revenues from nitrogen oxides and mining wastewaters must be used for the exclusive purposes of reducing nitrogen oxide emissions and saline-mining water discharges, respectively.

2.2 Facility Permits

Under Polish law all economic units are required to apply for permits for water intake and wastewater discharges, and emissions of air pollutants. Noise and vibration, disposal of solid waste, and treatment, storage and disposal of hazardous wastes also require permits. For all types of permits, similar application procedures are followed. The facility must apply for a permit and submit supporting documentation to the Voivod administrative authority. In practice, the Voivod Department of Environmental Protection will review the application and issue the permit. The critical document in the application package is the environmental impact statement which includes information on production levels, types of production processes utilized by the facility, fuels used, types and volumes of emissions or discharges that result from the proposed level of production, and the types of installed pollution controls. The applicant must also select an independent expert from an approved list compiled by the

Ministry of Environmental Protection to review the documentation and make a recommendation to the Voivod.

Upon submission of the documentation and the independent expert's report, the Voivod will make a determination of permitted emission or discharge levels. In the case of noise, a maximum noise level will be specified in the permit. In addition to the documentation submitted by the applicant, the Voivod is supposed to take into consideration current ambient pollution concentrations and determine the level of allowable pollution which will not result in degradation of ambient quality levels. In principle, permits may not be issued if the facility's proposed emissions would lead to violations of ambient standards. Voivod authorities may issue temporary permits if the facility submits a plan for implementing controls that would allow the facility to attain the allowable pollution levels. There is a strong incentive for facilities to obtain a temporary permit rather than to operate without a permit, since charges and fines (based on the proposed allowable level) are doubled for any pollutants not covered by a valid permit. New permits may also be denied if ambient standards are exceeded.

Pursuant to the EPL and the 1990 Ordinance of the Minister of EPNR&F, the Voivod is responsible for meeting the national ambient air quality standards, as specified by the Minister in Attachment 1 to this Ordinance. For air emissions, the permit applicant is also required to conduct air dispersion modelling to determine the contribution of facility emissions to ambient air quality. These findings also have to be approved by an independent expert. In its decision to issue the permit, the Voivod will attach considerable importance to the finding of the air dispersion modelling on the extent to which applicants' emissions contribute to violations of ambient standards.

An important exception to the practice of setting allowable levels to protect ambient environmental quality is the treatment of combustion sources. The 12 February 1990 Ordinance of the Minister of Environmental Protection, Natural Resources and Forestry on Air Protection introduced special, technology-based emission standards for SO_2, NO_x and particulates for major combustion sources with capacities greater than 200 kilowatts. All of these large combustion sources must comply with the technology standards by 1 January 1998.

Facility permits for water intake and discharge are separate but inter-related. They are issued together, but the permit for wastewater discharges requires an environmental impact statement, as required for air pollution permits. According to the Water Law, the wastewater discharge permit is required by all enterprises, and industrial and municipal sewage treatment plants which discharge sewage into surface waters or soils.

The applicant must submit a detailed environmental impact statement approved by an independent reviewer. The statement should include an enumeration of the types of water effluents disposed, their quantities and pollutant loads (concentrations of BOD, suspended solids and other pollutants). The permit specifies the allowable amount of the sewage that can be disposed, highest concentration of particular pollutants, and other technical features of sewage (for example, the radioactivity or temperature). Similar to air pollution permits, the Voivod takes into consideration the influence of applicants' wastewaters on the quality of surface waters. A general point of reference for surface-water quality is the classification system defined in the 5 November 1991 Ordinance of the Minister of EPNR&F. For wastewater discharge permits, the Voivod Environmental Protection Department will issue the permit with the concurrence of the Voivod Sanitary Inspectorate.

2.3 Enforcement and Noncompliance Sanctions

Under Polish law, the State Inspectorate for Environmental Protection has the lead responsibility for enforcement of permits. Inspectorate offices have been established in each of the 49 Voivods in Poland. While the Voivod Environmental Protection Department is responsible for issuing permits, the Inspectorate conducts facility inspections. The Inspectorate determines the extent of permit violations, levies noncompliance penalties, approves requests for deferral of penalties, and monitors enterprises' progress in addressing violations.

Noncompliance penalties are levied on the quantity of emissions or discharges in excess of the allowable limit specified in the facility permit. Penalty rates can be up to ten times the rate charged for charges. For wastewater, permits are specified in terms of pollutant concentrations in wastewater. A more complicated formula is used to determine the level of wastewater penalties. Nevertheless, penalties for wastewater are generally two to five times the applicable charge rate. Under Polish law, the *penalty rate* is fixed and cannot be reduced through negotiation between the enterprise and the State Inspectorate. However, the *degree* of the violation may be subject to negotiation. Continuous monitoring of emissions or discharges is not required. Thus, the extent of the violation must be estimated *ex post,* on the basis of the frequency of violations and calculations based on emission factors, level of production and use of inputs (for example, tons of coal burned by the enterprise).

For tax purposes, noncompliance penalties are treated differently from charges. Enterprises are not allowed to deduct noncompliance penalties as production expenses. Whereas a portion of the charge burden is shifted to other taxpayers, enterprises bear the full cost of penalties. Provisions have been made in the system of noncompliance penalties for penalized enterprises to defer the payment of penalties for 3–5 years. To qualify for this deferral, enterprises must implement measures to attain compliance status. If the enterprise is successful in attaining compliance, all or a portion of the deferred penalties may be waived, depending on the level of expenditures the enterprise makes to attain compliance. If these investments exceed the amount of accumulated penalties, the entire penalty is cancelled. If the enterprise fails to attain compliance, a 50 per cent surcharge of the original amount of the penalty is assessed and the deferred penalty plus the surcharge is collected. The Inspectorate can require the immediate payment of deferred fines if an enterprise fails to take actions (identifying technical options, applying for assistance, making investments) that would lead to compliance. In practice, however, regional inspectorates will not require payment of deferred charges until the deferral period ends, even if a polluter makes no effort to correct the violation (Warsaw Voivod Environmental Inspectorate, 1996). The Inspectorate is also empowered to shut down enterprises (or processes within facilities) which are chronically in violation of the terms of their permits or are operating without valid permits.

At the very beginning of the transformation process in Poland, the Ministry of Environmental Protection, Natural Resources and Forestry created the 'List of 80' worst polluters in Poland. The selection of enterprises for inclusion on the List of 80 was based on criteria including, *inter alia,* frequency of pollution discharges in excess of environmental standards, the degree of concentration of discharges, the location of enterprises and the range of negative impacts. The State Inspectorate prepared performance criteria for enterprises on the List of 80 related to reduction of the quantities of pollutants discharged by all of these enterprises and required listed enterprises to prepare implementation programmes describing the actions to be taken (for example, technological changes, restructuring of production process, construction and modernization of environmental controls), which would enable them to meet the established requirements. April 1993 was the date by which enterprises were expected to meet the performance criteria.

Approximately 800 enterprises, in addition to those on the List of 80, were also targeted as major polluters and subjected to special management supervision by the Voivod offices of the State Inspectorate. These enterprises were required to prepare environmental audits and develop compliance strategies.

An appeals procedure was also established to allow enterprises to contest penalties, suspension of deferred penalties, or orders to terminate facility operations. Enterprises may appeal such decisions to the Minister of Environmental Protection, Natural Resources and Forestry. Generally, if enterprises provide concrete proposals for addressing their noncompliance problems, they are successful in appealing orders to close down.

3   IMPLEMENTATION EXPERIENCE

To examine Poland's experiences in implementing its system of permitting, charges and fines, four topics will be discussed: (1) collection rates and rate calculations for charges and fines; (2) coverage of the system of facility permits; (3) capacity to monitor compliance and self-reported emission and discharge levels; and (4) success in eliminating noncompliance.

3.1   Collection Rates and Rate Calculations for Charges and Fines

As noted previously, Poland has increased charge and fine rates significantly in the 1990s compared to their levels in the 1980s, with most rates increasing 18–20 times in real terms. Prior to the large increase in charge rates, local authorities were able to collect nearly all of the charges and fines levied (see Table 9.2). Thus in 1990, 96.8 per cent of charges and fines levied were collected by Voivod authorities. In 1991, rates increased dramatically, leading to an increase in the amount of charges and fines levied from $31.1 million in 1990 to $523.1 million. However, authorities were able to collect only 73.7 per cent of imposed charges and fines. In 1992 and 1993 collection rates continued to decline as the total amount of charges and fines imposed increased. Data for 1994 indicate that collection rates declined slightly to 64.7 per cent from their 1993 level.

Table 9.3 provides data on charge and fine collections in 1994. As can be seen in the table, environmental charges represent the largest share of revenues (88.9 per cent), while charges from mineral extraction and fines are relatively minor sources of revenue. There are some striking differences between collection rates for charges. Collection rates for water withdrawals and air pollution are 96 and 90 per cent, respectively, while charges for waste disposal, wastewater and tree cutting are much lower. Because of the manner in which mineral extraction charges are collected, there is no assessment of the charge prior to collection. Fine collection rates are only 13 per cent, significantly lower in comparison to regular charges. However, the collection rate for fines is difficult to interpret because of provisions for deferral of penalties for three to five years. Most of the uncollected fines are deferred and subsequently subtracted from investment costs incurred to achieve compliance. It should also be noted

Table 9.2 Environmental charges and fines (in millions of US$)

	1990	1991	1992	1993
Imposed charges and fines	31.1	523.1	649.6	660.6
Amounts actually collected	30.1	385.7	446.8	428.5
Collection efficiency rate (%)	96.8	73.7	68.8	64.9

Source: Kruszewski (1994).

that charge payments for a given year include revenues for charges imposed during the current year plus revenues for delinquent payments from previous years. Thus, year-to-year fluctuations in collection rates may be attributed partly to the payment of delinquent charges and unpaid charges in the current year.

As noted earlier, the level of charges imposed is based on self-reported emission and discharge amounts. The ability of regional inspectorates to verify these amounts is limited by staff resources and compounded by the large number of facilities and individual pollutants that must be checked. In addition, there appear to be a number of small and medium-sized enterprises that have not been included in the charge system. Similar problems have been noted earlier for fines. Thus, it appears that potential

Table 9.3 Environmental charge and fine revenues in 1994

Type of charge	Charges assessed (US$m.)	Charges collected (US$m.)	Collection rate (%)
Environmental charges	585.9	431.6	74
Water withdrawals	56.9	54.6	96
Wastewater discharges	200.9	91.3	45
Air pollution	245.8	222.0	90
Waste disposal	81.9	63.5	77
Tree cutting	0.2	0.1	47
Noncompliance fines	101.4	13.5	13
Mineral extraction charges	n.a.	41.2	n.a

Source: NFOSiGW (1995).

charges and fines are probably higher than levels reported.

While some effort has been made in Poland to relate charge and fine rates to environmental damages and/or compliance costs, there is no transparent methodology applied at the present time. Generally, there is better justification for differentiated rates based on relative toxicity than for the absolute levels of charges. However, there are still some major flaws in rates for different users, particularly for water and wastewater. For example, charges for use of surface- and groundwater are as much as 31 and 47 times higher for certain classes of industrial users than for households, municipalities and agricultural interests (Zylicz, 1994). Similar distortions are observed for charges on wastewater discharges, although the spread is more on the order of ten times. Currently, a number of environmental economists and environmentalists are recommending that wastewater discharge rates be differentiated according to the quality of watercourses into which a facility discharges its effluent.

3.2   Coverage of the System of Facility Permits

Most of the larger facilities in Poland operate under either a valid permit or a temporary permit. In 1992, 17,389 facilities were registered as water polluters and 46,305 as air polluters (Broniewicz et al., 1994, p. 3). Although registered facilities are required to apply for facility permits, it is estimated that nearly half of facilities operate without valid permits (Broniewicz et al., 1994, pp. 5–6). The backlog is largely attributable to limited local resources to process permit applications. In addition, there is growing concern about the permitting requirements resulting from Poland's harmonization with environmental directives of the European Union. In many instances, current facility permits will need to be revised. When added to the resource and staff costs to prepare permits for those facilities which currently have no permits, local environmental authorities will face an enormous challenge.

The difficulty of updating and preparing new permits could also be affected by additional factors. First, some fine tuning of national ambient standards might require revisions of airshed modelling and modification of allowable emission and discharge levels. Second, as the State Inspectorate proceeds with plans to improve the system of ambient monitoring, new data may also raise some questions about the quality of airshed monitoring and the ability of Voivods to meet ambient standards. Third, assuming the new Water Law is enacted, there will be a shift in responsibility for protecting the quality of Poland's ground and surface waters from the Voivods to seven regional water authorities. This could lead to much greater consideration of upstream loading and downstream water quality than under current Voivod management.

Finally, there is a need to upgrade the skills of environmental officials and staff at the regional and local levels. The terms of the permits given by different regional administration bodies differ fairly significantly. These disparities cannot be attributed only to differences in climatic and topographical features, but also to considerable differences in professional skills and experience of environmental inspectors and other civil servants being employed in the environmental protection departments.

3.3 Capacity to Monitor Compliance and Verify Pollution Levels

The Law on the State Inspectorate for Environmental Protection of 20 July 1991 provided the basis for Poland's monitoring system. Polish authorities began to develop the extensive environmental monitoring network for air, noise, surface and ground waters, nature protection and radioactive contamination in 1992. The system is expected to be completed by 1997. Once the monitoring system is fully operational, it is anticipated that new monitoring data will play a greater role in the development of environmental policy and provide a basis to revise ambient quality standards. In addition to the development of this extensive monitoring network, regional inspectorates have been created with responsibilities for monitoring compliance with facility permits and for verifying the accuracy of pollution levels reported by facilities (for calculation of charges).

The efforts of Voivod inspectorates to carry out these functions are hampered by a number of factors. First, there is no explicit requirement for permitted facilities to undertake an ongoing monitoring programme and to invest in on-site monitoring equipment. This means that the historical record must be reconstructed from other related and known data, such as production levels, processes and technologies. Second, Voivods lack the staff resources to conduct frequent inspections of facilities to verify compliance with permit requirements and calculate pollutant levels. Third, because fines are significantly higher than charges, there are greater incentives to avoid fines. Noncomplying firms can contest the extent of violations of permit terms (assuming they are detected at all). In addition, once a fine is issued, firms can request deferrals of fines while they attempt to implement investment projects to correct violations. While the deferral policy most often leads to elimination of violations, it nevertheless increases the workload of the regional inspectorates; monitoring of a polluter's plan to correct a permit violation is undertaken by the regional inspectorate. The added penalty for failing to comply after requesting deferral is only a one-time charge of 50 per cent. For a three-to five-year deferral this is a small charge when one considers that recent inflation rates have been 25–35 per cent per year.

3.4 Success in Eliminating Noncompliance

The major policy initiatives undertaken in Poland to achieve compliance with environmental regulations have been noted earlier and include charges, fines, subsidized financing available from the National and Voivod environmental funds, and targeted noncompliance monitoring of enterprises on the List of 80. It is difficult to examine the effect that charges have on compliance decisions, independent of the subsidized financing. As will be discussed in greater detail in Section 4, charge rates are not high enough to provide incentives for investments. However, payment of charges generates substantial revenues, which are recirculated as grants and soft loans to enable enterprises to finance environmental investments. These recirculated revenues provide almost half of the funds for environmental investment in Poland. Noncompliance fine rates are quite significant and clearly have an impact on violators. A large proportion of violators request deferral of fines, thereby committing themselves to programmes of investments. According to the OECD, approximately 70 per cent of violators are now in compliance with permit requirements (OECD, 1995, p. 100).

The List of 80 and similar Voivodship lists based on the same concept which include almost 800 enterprises have now existed for almost five years. Seven enterprises from the List of 80 have been permanently closed and production at another 22 facilities was at least partially or temporarily interrupted. The preparation of the list has helped central and local administrations focus their attentions on the environmental hot spots in Poland and stimulated communities in the neighbourhoods of listed enterprises to take an interest in solving environmental problems. The actions of enterprises under public pressure and scrutiny have brought considerable reductions of pollutants emitted by enterprises on the List of 80. According to a report of the State Environmental Inspectorate conducted in 1993, the following environmental improvements have been made by listed enterprises (relative to 1989 levels): (1) emissions of particulate matter decreased by about 67 per cent; (2) emissions of gases decreased by about 44 per cent; (3) discharged sewage decreased by about 37 per cent; and (4) stored waste decreased by about 42 per cent.

At least a portion of these reductions is attributable to the decrease in economic output of these enterprises during 1990–93, but the majority of these reductions are generally attributed to actions taken by enterprises to reduce emissions. In May 1994, several enterprises from the List of 80 were removed and a few new ones were listed. At present, 74 enterprises are on the list. The State Environmental Inspectorate projects that six more enterprises may be removed from the list by the end of 1995 and an additional ten enterprises by the end 1996.

4 THE STRUCTURE OF INCENTIVES IN THE POLISH SYSTEM OF ENVIRONMENTAL CHARGES AND PERMITS

There is considerable support among economists for the use of charges to provide incentives for polluters to make decisions on investments in pollution control. Presumably, polluters will undertake those investments for which the marginal cost of control is less than the charge rate. This shifts the burden of achieving environmental goals to the policy maker, who must establish charge rates at levels that will encourage polluters to make reductions sufficient to meet environmental targets. If rates are set too low, too many polluters will prefer to pay the charges rather than invest in controls. If the rates are two high, assuming environmental agencies are able to collect the high charges, polluters will *overcontrol*.

Environmental charges in Poland are among the highest in the world. Yet even at these levels, charges are not high enough to stimulate polluters to make investments adequate to meet emission or discharge targets. For example, the charge on SO_2 is approximately \$75 per ton. However, the estimated marginal cost of investments to achieve a 30 per cent reduction (equivalent to the new requirements that take effect in 1998) in SO_2 using the RAINS (Regional Acidification Information and Simulation) model developed by the International Institute for Applied Science Analysis, is approximately \$600 per ton for large combustion sources. There is at least some anecdotal evidence that the high charge rates have provided incentives for polluters to make low-cost improvements to reduce emissions of particulates and sulphur dioxide, although not necessarily sufficient to generate compliance with the standards.

Overall, the system of charges plays a secondary role in providing incentives for pollution reduction. Charges are high enough in Poland to encourage facilities to make low-cost and win–win investments or to make improvements in 'housekeeping' practices. Charges generally do not encourage polluters to exceed permit requirements by a substantial amount. Thus, where stricter standards are phased in (for example, the 1998 major source combustion standards), charges rates are not high enough to encourage early or accelerated compliance. More traditional enforcement measures including noncompliance fines and legal authority to close down a chronic violator provide incentives for meeting current requirements.

It might be possible to increase charge rates above their present level to provide greater incentives for pollution-control measures, but there does not appear to be political support to increase charge rates. Even if charge rates were increased, there is considerable concern among economists in Poland that enterprises lack the expertise to evaluate abatement costs and determine least-cost solutions to minimize the sum of charge payments and abatement costs. While capital costs of environmental investments are known, enterprises have limited experience in estimating depreciation or 'running' costs (O & M). Partly, this limited experience can be attributed to the Polish Law on Accounting, which does not require enterprises to create separate accounts for environmental protection. In addition, official statistics only cover capital expenditures on environmental investment. Furthermore, environmental protection expenditures are defined *sensu stricto* as end-of-pipe, stack and waste-disposal investments and do not include investments in energy and resource efficiency, production process improvements, or waste minimization. Even if charge rates were higher, it is therefore not clear that enterprises would be able to respond to these 'price signals' and implement least-cost solutions.

Charges and fines generate more than $400 million per year and these revenues are recirculated by the environmental funds for environmental investments and other environmental activities. All new revenues from charges and fines are earmarked for expenditures to address the types of pollution from which the charges were generated. The standard concern with earmarking is that these revenues do not necessarily achieve the greatest net social benefits. The most efficient method of distributing charge revenues would be to allocate them to projects generating the greatest net social benefits, whether for air, water or waste. For sulphur dioxide and nitrogen oxide, the recirculation distortion is somewhat less because the charges, while earmarked for reducing these pollutants, can be distributed for abatement projects in any region and any enterprise. For example, charges collected from a power plant do not have to be spent on expensive end-of-pipe treatment, but may be used to co-finance a coal-washing project elsewhere in Poland.

Much more severe distortions in the efficient use of charge revenues are caused by the fact that on balance Poland's environmental funds have transferred resources from manufacturing industries to municipalities. This practice reinforces and, in the case of charges on water, amplifies the wrong signals sent to the household sector and municipalities. This is yet another example of how equity concerns interfere with the efficiency of the charge mechanism. In addition to these problems, the National Fund and Volvid environmental funds have – until recently, and then only for a selected number of funds – employed poor project screening and assessment procedures in evaluating grants and loans. These practices sometimes lead to the selection of projects that are not cost effective even within their respective category of abatement (such as sulphur removal, municipal waste abatement and so on).

Given the existence of earmarking and the current project selection practices, it is interesting to look more closely at the largest industrial polluters and to speculate whether they are better served by the current system of charges (and recirculated grants and loans), compared to an alternative structure where charges are retained by firms and invested in environmental controls. In a sample of 112 of the largest polluters in Poland, Broniewicz et al. (1994) estimated the percentage of enterprise expenditures devoted to charge payments, abatement costs and other pollution costs. For the entire sample, pollution charges in 1992 represented 4.9 per cent of expenditures, abatement costs were 1.6 per cent and other pollution costs were 0.5 per cent. Total expenditures on charges for the surveyed enterprises were $190 million compared to investments of only $108 million.

Interestingly, while environmental funds provided about half of the financing for environmental investments, grants and loans financed by environmental funds accounted for only 16 per cent of investment expenditures among surveyed enterprises. On the surface, it would seem that these large polluters would be better off if they could retain charge payments, but there may be other explanations for the limited use of grants and loans by surveyed enterprises; investments not identified or applications not prepared for submission to funds, other more pressing non-environmental investments given priority over environmental investments, and so on. In addition, to the extent that investment proposals are evaluated by funds using criteria related to cost effectiveness or net benefits achieved, charge revenues may achieve greater net benefits when invested by funds than if enterprises are allowed to retain their own charges for investments. Retention of charges would also require an additional level of monitoring by environmental authorities to ensure that retained charges are actually used for environmental purposes.

5 RECOMMENDATIONS FOR IMPROVING THE SYSTEM OF ENVIRONMENTAL CHARGES

While Poland has established a comprehensive and integrated system of charges, fines and permitting, there are a small number of problems that, if addressed, would improve the system's performance.

1. *Fine-tuning of charge rates*   Under the current system there are too many pollutants with separate charges. As noted, it is difficult for local authorities to monitor actual emissions or discharges. There are also concerns about the absolute rates of charges and the relative differences between charge rates for pollutants of varying toxicities. Third, the justification for differential charge rates for different classes of sources is very weak, and removing distortions in charge rates differentiated by toxicity or user would enhance the credibility of the charge system. It has also been suggested that the system of charges be streamlined and simplified to enhance monitoring and collection.

2. *Introduction of product charges*   When charges severely affect firms' profits, they may respond by refusing to pay or underreporting emissions rather than investing in abatement or closing down. Social problems during the transition period and barriers to financing of environmental investments may also contribute to evasion of charges. The experience of OECD countries suggests that product charges may be a more effective revenue-raising instrument than pollution charges. Fuel charges were first proposed by the MEPNR&F in Poland in 1990, but did not garner adequate support in the Council of Ministers for consideration by parliament. Currently, the Ministry of Environmental Protection is preparing a proposal for new product and deposit-refund charges. The proposal is still at the conceptual level, but if it receives the support of parliament's Environmental Commission, it may be submitted for legislative consideration in 1996. Revenues from product charges would go to the National Fund, allowing some reallocation of charge revenues from the National Fund to Voivod and Gmina funds.

3. *Clarify responsibilities for meeting ambient quality standards*   At the current time, there is often ambiguity and overlapping authority for meeting ambient quality standards. Much of the problem stems from conflicts between Voivods, municipalities and Gminas over jurisdiction. Permits are issued by the Voivod, but municipalities play a role in developing environmental master plans for their communities and receive a portion of charge and fine revenues from Gmina funds. Poland is now developing a new organic or framework environmental law. This development provides an opportunity to clarify responsibilities for meeting environmental quality standards among different levels of government.

   The situation for water pollution is particularly unclear. Poland is in the process of creating seven regional water districts delineated according to the major river basins in the country. However, the new water law which would transfer authority from Voivods to these regional districts is not expected to be enacted until the issue of how to distribute water and wastewater charges is resolved to the satisfaction of Voivods. If all of these charges are given to the new regional water funds, Voivod funds would lose almost half of their current working capital. From an efficiency perspective, regional water funds would be best able to promote environmental improvements in the river basins. However, Voivods would lose control of resources for local wastewater projects (unless they received support from regional water districts).

4. *Increased flexibility in permits*   At the current time, facility permits do not accommodate the use of economic instruments such as offsets, bubbles or emissions trading. Permits specify emissions and discharges for each individual process and emission or discharge source. With the exception of the new major source combustion standards, facilities are not allowed to average across sources. Some recent work on air pollution suggests there may be substantial benefits from the use of these economic instruments. In the Voivod of Opole, a pilot emissions trading programme in SO_2 is being developed. However, changes in Polish permitting laws will be required to implement the pilot programme.

REFERENCES

Broniewicz, E., B. Poskrobko and T. Zylicz (1994), 'Internalizing Environmental Impacts of Industry in Poland: Preliminary Empirical Evidence', Paper presented at the 5th Annual Conference of the European Association of Environmental and Resource Economists, Ireland.

Kruszewski, J. (1994), 'Poland', in P. Francis (ed.), *National Environmental Protection Funds in Central and Eastern Europe. Case Studies of Bulgaria, the Czech Republic, Hungary, Poland and the Slovak Republic*, Budapest: Regional Environmental Centre for Central and and Eastern Europe.

National Fund for Environmental Protection and Water Management (NFOSiGW) (1995), *The Report of the National Fund for the Year 1994*, Warsaw.

Spyrka J. (1994), 'Poland. Part B', in J. Klarer (ed.), *Use of Economic Instruments in Environmental Protection in Central and Eastern Europe. Case Studies of Bulgaria, the Czech Republic, Hungary, Poland, Romania, the Slovak Republic, and Slovenia*, Budapest: Regional Environmental Centre for Central and Eastern Europe.

Warsaw Voivod Environmental Inspectorate (1996), Personal Communication with Director, January.

Zylicz, T. (1994), 'Taxation and Environment in Poland', in *Taxation and Environment in European Economies in Transition*. Paris: OECD.

APPENDIX 9A

Table 9A.1   Charges for emissions for atmospheric pollutants in Poland in 1995

	US$/ton
Atmospheric pollutant	
Acrylonitryl (aerosol), asbestos, benzene, benzo-a-pyrene, chlorinated vinyl (gaseous)	$53,739.00
Arsenic, chromium, nickel (fee rates apply to metal content)	$53,739.00
Bismuth, cerium, tin, zinc, cadmium, cobalt, manganese, mercury, molybdenum, lead (fee rates apply to metal content)	$26,869.00
Chlorofluorocarbon compounds, carbon tetrachloride, dioxin, halons, polychlorinated biphenyls, 1,1,1 trichloroethane	$26,869.00
Heterocyclic compounds	$1,817.39
Nitric, nitrous, and related compounds	$700.00
Amines, cyclic and aromatic alcohols, organic and elemental sulphur	$360.87
Organic acids and related compounds	$291.30
Carbon bisulphide	$252.17
Particulates (cement, silicate, fertilizer, solvent, carbon-graphite, carbon black), ketones and inorganic acids, aromatic and cyclic aldehydes, ether, alipathic and related alcohols, isocyclic compounds, aromatic and cyclic hydrocarbons, non-metallic elements, nonmetallic salts, non-metallic oxides	$213.04
Aliphatic aldehydes and related compounds	$143.48
Sulphur dioxide, nitrogen oxides	$82.61
Ammonia, HCFCs, other halons, particulates (polymers and lignite), oils	$73.91
Particulates from fuel combustion, all other particulates	$43.48
Carbon monoxide, hydrocarbons	$21.74
Carbon dioxide, methane (fee per ton)	$0.04
Charges for evaporative emissions	
Filling fuel storage tanks with fixed roof	$0.83
Filling fuel storage tanks with floating roof	$0.05
Filling underground and above ground storage tanks	$0.46
Filling train fuel tanks	$0.46
Filling car/truck tanks	$0.35
Filling automobile tanks	$0.51

Table 9A.2 Waste-disposal charges in Poland in 1995

Type of waste	US$/ton

Group I: The most toxic substances, for example:
  –waste with mercury or its non-organic compounds
    (except for HgS) with mercury content of more than
    0.005%
  –waste with arsenic compounds (except sulphides)
    content of more than 0.005%
  –waste with selenium content of more than 0.005%
  –used oils and greases
  –asbestos waste
	$21.49

Group II:
  –waste with fluorine compounds less than 0.5%
  –waste with mercury and its compounds (except for
    HgS) in concentration 0.005–0.001%
  –waste with arsenic or its compounds (except those
    in Group I)
  –banned agricultural chemicals
  –used catalytic converters
	$8.06

Group III:
  –waste from the sodium industry
  –used adsorbents (for example activated carbon)
  –mineral cement-calcium dusts
  –asbestos and cement-asbestos waste
  –waste coming from vessels and harbours
	$2.69

Group IV:
  –waste resulting from removing sulphur from fumes
  –sludge after drinking water treatment
  –waste coming from textile industry
  –waste coming from mines
  –paper, glass
	$1.61

Table 9A.3   Charges for wastewater and saline (coalmining) water in Poland in 1995 ($/ton)*

	BOD$_5$	COD	SS	C&S	HM	VP
Power generation, fuel processing, chemical, metallurgical, machine and light industries	1,722	1,206	74	6	8,600	3,226
Pulp and paper industries	732	434	74	6	8,600	3,226
Food industries	430	290	74	6	8,600	3,226
Municipal sewage, hospitals and social care institutions	172	96	74	6	8,600	226
Other (except saline coal-mining waters)	861	483	74	6	8,600	3,226
Saline coal-mining waters discharged directly to an aquifer	–	–	–	48	8,600	–
Saline coal-mining waters discharged from dosing reservoirs	–	–	–	6	8,600	–

Notes:
* For discharge into lakes and retention reservoirs double rates apply. Charges applied in Katowice region are twice as high as in the table.
BOD$_5$ = Biochemical oxygen demand during the first five days.
COD   = Chemical oxygen demand.
SS   = Suspended solids.
C&S   = Chloride and sulphate ions.
HM   = Heavy metals.
VP   = Volatile phenols.
The total payment is calculated as the maximum charge for BOD$_5$, COD, SS and C&S; added to the charge for HM and VP, or Max (BOD$_5$, COD, SS, and C&S) + HM + VP.

10. Implementation of Pollution Charge Systems in a Transition Economy: The Case of Slovakia

Thomas H. Owen, Jozef Myjavec and Danka Jassikova

1 INTRODUCTION

Slovakia is a relatively new country, becoming independent on 1 January 1993. It is situated in the heart of Central Europe. Its more than 5 million inhabitants occupy 49,036 km², including 24,471 km² of agricultural land, 14,860 km² of arable land, and 19,911 km² of forest lands. The area covered by water is 940 km². Under the Slovak Constitution, every citizen has the right to a satisfactory environment, a duty to protect and improve the environmental and cultural heritage and the right to timely and complete information about the state of the environment. The government has the responsibility for the efficient use of natural resources, the maintenance of ecological balance and the protection of the environment. This constitutional framework is the basis of the environmental management and pollution charge systems in Slovakia.

Led by declining industrial production, largely due to the transitional state of the economy, during 1992–94 GNP declined by as much as 10 per cent per year accompanied by significant inflation. Recently this has changed and in 1995 GNP increased by 4 per cent. Annual inflation was also reduced by half (Ministry of Environment, 1995).

The Slovak government has adopted a national environmental policy which recognizes the impact of the environment on life expectancy and public health.[1] It acknowledges the deterioration of the environment created by outdated technology and infrastructure and particularly cites

---

1 The Ministry of Environment has principal responsibility for issues concerning the environment, but several other ministries also have related responsibilities.

problems associated with an energy and raw material-intensive economy. The policy document identifies specific environmental issues which differ by region, and also notes that in the field of air pollution, in spite of recent reductions, Slovakia still produces four times the sulphur dioxide (SO_2) of neighbouring Austria (National Council, 1993). This is with a 1993 GNP almost 18 times lower (World Bank, 1995). With respect to water, Slovak water reserves are slightly less than 60 per cent of the European average, while daily consumption of drinking water is double that of Austria.

Although more treatment plants are being constructed, the discharge of untreated waters continues to be a problem. Seventy-five per cent of the economically important rivers are categorized as fourth- or fifth-class purities.[2] Groundwater has also significantly deteriorated over time. In 1991, 87 per cent of the samples collected were judged unsuitable for consumption, which is greater than the 63 per cent figure noted in 1983. Waste is also a major problem, with only 335 facilities licensed out of 8,372 known landfills (National Council, 1993).

The Ministry of Environment was reorganized in 1994 and this reorganization created a new Division of Environmental Economy, which is responsible for a variety of topics, including implementation of economic instruments. Economic instruments are an important component of the overall environmental policy framework in Slovakia. They are mandated by legislation and based upon pollution limits specified through a permitting process. In addition to pollution charges, there are penalties for breaches of regulations. Differential rates and other tax measures designed to improve environmental behaviour have also been implemented in Slovakia.[3]

The polluter pays principle is implemented through a system of charges which apply to most wastes released into the environment. The charges and penalties provide revenues for the State Environmental Fund, which are used as supplementary sources of financing for environmental projects.[4] In 1993, about 70 per cent of the State Environmental Fund receipts came from pollution charges and penalties, with the balance coming primarily from the state budget (CowiConsult, 1995).

This chapter provides an overview and analysis of the permitting system, describes the system of environmental charges and penalties

2   These are the most polluted rivers.
3   There are reduced levels of VAT on certain environmentally-friendly products (for example, electric cars and cars with catalytic converters are tax free), a one-year exemption from income tax is available for alternative energy sources, such as biogas, solar and geothermal energy plants, and there is an exemption from property tax for owners of natural protected areas.
4   There are some exceptions to this rule. For example, the revenues collected from user charges for the withdrawal of surface and groundwater go to River Basin Enterprises, and charges collected for air pollution emissions from small sources go to municipal budgets.

currently in force in Slovakia, and evaluates the impact and efficiency of these charges. However, several points of caution should be made at the outset. First, in Slovakia, as in other economies in transition, recession, structural changes in the economy and the impact of privatization are overriding issues, which directly affect the impact and efficiency of the pollution charge system. Second, evaluation of environmental charges is complicated by many factors not present in most other countries that rely on such instruments. For example, the state maintains a large ownership share in key enterprises, so some pollution charges represent transfers from one state account to another. Pollution-control investments are also largely financed from state funds, with only a small share paid by the affected enterprises.

A large share of pollution charges also go unpaid – only about one-third of amounts due have been collected in the past two years. Finally, output prices are sometimes determined in whole or part by the state, potentially limiting the ability of a firm to increase prices in response to an increase in operating costs. These facts, combined with the limited Slovak experience in applying economic tools and substantial shortage of qualified environmental economists, make it very difficult to analyse and understand incentive effects, let alone fully implement and support an effective systematic approach to the use of economic instruments.

2 THE PERMITTING PROCESS

Assessing charges, penalties and enforcing breaches of any provision of permits are the responsibilities of environmental offices which are supervised directly by the ministry. Permits are issued by 38 District (*Okresny urad*) and 121 Subdistrict (*Obvodny urad*) environmental offices. These offices also inspect facilities, order remedial actions and assess penalties. In general, they issue the orders for steps to be taken by polluters to be in compliance with environmental laws. Jurisdiction is normally determined by the catchment area in which environmental impacts are expected to occur. For example, a ministry decision would be required for activities of national or international scope, whereas a local environmental office decision would be made with respect to a facility having only local impacts. There is also an established procedure for the state administration to reach these decisions, including requirements to promptly notify affected parties (National Council, 1967).

Municipal authorities (*obecne urady*) play only a very limited role in the field of environmental management. Plans exist, however, to transfer certain powers and responsibilities to self-governing municipalities, and the Slovak government intends to integrate the general and special offices of state administration at the district and subdistrict levels. The Slovak Environmental Inspectorate, which includes both a central component and seven regional offices, monitors compliance with decisions given by the environmental offices. The Inspectorate may also impose penalties (National Council, 1990).

3 POLLUTION CHARGES

There are two types of pollution charges. A basic charge is due from each polluter based upon formulas for individual pollutants as provided for in the legislation. In some cases, an additional charge is applied to discharges over and above those provided for in the permit. For water pollutants, charges are determined based on estimates of the potential damage that may be caused by pollutants, and are calculated from the mass of the pollutant and the volume of the receiving waters.

For air pollutants, charges have one part set by legislation (which provides for uniform standards throughout Slovakia) and the second is a 50 per cent surcharge on quantities that exceed source-level emission limits. The first charge is deductible as an expense for income tax purposes, but the surcharge is paid out of after-tax profits.

Air pollution charges are calculated by environmental offices based upon previous years' annual reports. The Environmental Fund receives a copy of the decisions and payments are made directly to the Environment Fund. The Fund has a computerized tracking system and notifies the responsible environmental office when payments are not made. In the case of water, monitoring and enforcement are delegated to river basin authorities who calculate charges and collect payments. The money is then forwarded to the Environment Fund on a monthly basis. In the field of waste management, operators have the responsibility for establishment and collection of charges.

Penalties are provided for in the legislation and deal with infringements of the law. For example, discharging pollutants without required permits or breaching terms of a permit by emitting greater volumes or higher concentrations than stipulated, draws penalties of various sizes depending on violations.[5] Revenues from penalties go to the Slovak Environmental Fund (National Council, 1991).

---

5   For example, for breach of provisions in the Act on Air Protection, penalties range from SK 5,000 to SK 10 million depending on the source and the seriousness of the violation.

During the period of economic transition, the Ministry of Environment has phased-in charges so that polluters have time to adjust to these new costs. Polluters can then adopt alternate technologies, engage in input substitution or waste-minimization activities, and therefore manage their pollution charge liabilities. This gradual approach, in principle, allows a softening of the full impact of charges.

3.1 Charges for Air Pollution

Only medium and large sources are included in the national system of pollution charges. Large sources are defined as thermal units above 50 MW, as well as other industries such as coke, steel, glass, iron, cement and heavy chemical production. Medium sources are thermal units within the range of 0.2 MW to 50 MW and other technologies defined in Act No. 309/1991.

A charge is imposed for every ton of discharge emitted from these sources; based upon obligatory self-measurement. Revenues from these charges amounted to SK 238 million in 1993 and SK 95 million in 1992 (Klaren (ed.), 1994). The charges seek to provide economic incentives for air polluters to take pollution reduction measures. As was indicated earlier, another function is to generate income and to provide financial support for environmental investments through the Slovak Environmental Fund.

Charges are imposed on five categories of air pollutant emissions determined by Ministry Decree 407/1992. The decree specifies both a final rate and a schedule to meet the final rate. The main pollutants are solid particles, SO_2 NO_x and CO. Approximately 120 other pollutants are divided into four groups based on toxicities. Charge rates are established according to these groupings (National Council, 1995). However, the level of charges does not reflect the cost of elimination of these emissions into the air. Final rates, as well as the phase-in schedule for the period 1992–98 are given in Table 10.1.

Environmental offices annually review the records presented by firms and assess the charges. There are also random inspections by environmental offices or Inspectorates. However, the level of collection of imposed charges is a problem (approximately two-thirds of those charged are insolvent) and enforcement is only possible through the courts. This is a timely and cumbersome process and alternatives such as direct deduction from polluters' bank accounts are not available.

Charges of up to SK 10,000 per source per year may be imposed on operators of small pollution sources (that is, those not exceeding 0.2 MW) by the municipalities in which sources are located, but collections are low.

Table 10.1   Final charge rates for emissions into air (SK/ton)

	SK/ton
Main pollutants	
Solid emissions (particulates)	3,000
SO_2	1,000
NO_x	800
CO	600
Other pollutants (includes approximately 120 pollutants)	
1st class (includes asbestos, Cd, Hg, benzopyrene)	20,000
2nd class (includes As and its compounds, Pb, Zn, CN)	10,000
3rd class (includes formaldehyde, Cl, H_2S, HF, phenol)	5,000
4th class (acetone, ammonia, Cl-benzene, styrene, HCl)	1,000
Air pollution charges are gradually being increased	
1992	20% of the final rate
1993	40%
1994–95	60%
1996–97	80%
after 1998	100%

Note: In SK/ton (US$1 = SK 30).

Except for a few large cities and areas of significant industrialization, small-source charges are seldom enforced (Klaren (ed.), 1994).

3.2   Charges for Water Pollution

This charge is integrated with the permit system and, as is the case for air pollution, is levied based on self-monitoring by polluters. The amounts of charges depend on the quantity of pollutants in the wastewater and on the quantity[6] of receiving waters. Charges are collected by river basin authorities, which are state enterprises. In 1993, SK 293 million in effluent charges were collected. In 1992, collections totalled SK 405 million (Klarer (ed.), 1994).

The charges attempt to stimulate water polluters to treat water prior to discharging it and also ensure that those with wastewater treatment

---

6   But not on the quality.

plants are not put at a competitive disadvantage *vis-à-vis* those without plants. Basic charges are levied on BOD_5, insoluble substances and crude oil substances, alkalinity and acidity, and dissolved inorganic salts (Government of Czechoslovakia, 1989). Additional charges of up to 200 per cent of base rates may be levied to reflect high levels of damage to receiving waters. These additional charges must be paid from after-tax profits.

There also exist charges for sewerage which are levied by individual municipalities and therefore can be partially subsidized if a municipality so chooses. The average consumption of water per person in 1992 was about 178 litres per day and about 2.7 million Slovak citizens had sewer connections. The level of charges for sewerage was constant until 1990, but since 1991 the charges have been significantly increased. They cover all costs, including the costs of releasing wastewater into surface waters. Since 1993, charge rates have been set to allow full cost recovery by municipalities, and charges are usually included in the cost of drinking water. Maximum rates are as follows:

water rate   4.00 SK/m^3
sewerage   3.00 SK/m^3.

Defining Z as the amount of pollution subject to charges in tons per year, total basic charge payments (in thousands of SK/year) are given in Table 10.2.

4   CONCLUSION

Economic instruments for air and water protection were introduced during the period when there was a centrally-planned economy. In some cases this resulted in problems as Slovakia moved towards a market system. For example, in the water field approximately 300 exemptions from the law were given in 1990. In addition, pricing policy did not reflect the true costs of water, fuel, heat, energy and raw materials. In spite of such problems, new legislation is being prepared which expands the use of economic tools, including the increasingly important consideration of harmonization with the EU. There is also a growing realization that the system of economic tools, including fees and charges must be revised.

Several factors seem to prevent the pollution charge system from fully accomplishing the objective to make polluters pay the full costs of their actions. First, of the nominal amounts due under the charge system about two-thirds are uncollected. While measures are being taken to improve

Table 10.2   Total effluent charge payments for various water pollutants (thousands of SK/year)

BOD_5
Total charges = $21.5 \times Z^{0.8265}$

Insoluble substances
Total charges = $2.34 \times Z^{0.7514}$

Crude oil substances
Total charges = $Z \times K$
K = Per unit rate depending on concentration of pollutants

5 to 10 mg/l	= 1.00 SK/m³
10 to 20 mg/l	= 1.50 SKm³
20 to 35 mg/l	= 2.00 SK/m³
35 to 50 mg/l	= 2.50 SK/m³
more than 50 mg/l	= 3.00 SK/m³

Alkalinity and acidity
Total charge = $Z \times M$
M varies depending on the level of alkalinity or acidity at the rate of 135 SK/kmol of alkalinity/acidity

Dissolved inorganic salts
Total charge = $Z \times S$
S = Per unit rate depending on concentration of pollutants

600 SK/t	with flow up to	0.01 m³/s
300 SK/t		0.01–0.1 m³/s
200 SK/t		0.1–1.0 m³/s
150 SK/t		1.0–10.0 m³/s
120 SK/t	more than	10.0 m³/s

---

collection rates, difficult economic circumstances remain a problem for many firms.

There has also been no attempt to determine whether charges have been set at levels reflecting the damage which is the real cost of pollution, and in most cases charges appear to be much less than the incremental cost of control, suggesting that incentive effects are limited. The main problem with evaluating the impact of the system is the same one which faced those designing the system, namely the lack of economic data directly collected in the Slovak Republic. Furthermore, the data that are reported reflect a number of interactions between the state and the polluting enterprises, interactions that make it difficult to assess the significance of what is reported. For example, pollution abatement costs are heavily subsidized in the Slovak Republic, with firms paying on average only 16 per cent of the total cost of projects.

No systematic review has therefore been made of the impact of fees and charges, either at the level of the individual enterprise or for the economy generally. The Ministry of Environment is generally accelerating the application of economic analysis to environmental policy and decision making.[7] It therefore may be possible to conduct the necessary analyses in the future.

In the meantime, there is some evidence, though not directly attributable to the pollution charge system, suggesting that the environmental situation in Slovakia is improving. A recent report prepared for the OECD, for example, pointed out: 'in the past five years Slovakia has experienced a substantial decline in pollution levels exceeding the decline in GDP. This indicates that the decrease in pollution levels is not merely a result of output declines' (CowiConsult, 1995 p.). Though there is a general lack of capital for environmental investments in Slovakia, it is also known that some major industrial enterprises have initiated ambitious environmental investment programmes. For example, Slovnaft, the major refinery, has announced a SK 15 billion (US$500 million) programme over three years.[8]

Slovakia is still in the process of implementing a systematic approach to the use of charges. As is the case for any evolving system, there will be startup challenges and adjustments. Although it was not discussed in this chapter, charges for waste management provide a useful example. The charge for waste disposal is a relatively new charge and revenues have been lower than anticipated. A number of factors may be contributing to this situation, including inaccurate estimates of volume and characteristics of waste streams, system startup problems, or possible disposal outside of the organized system. The charges and monitoring will increase

7 Recently the Ministry of Environment commissioned an analytical study of the Economic Implications of the Implementation in Slovakia of the Copenhagen Annex to the Montreal Protocol.

8 Slovnaft is undertaking the most difficult project in its history. The APPOLLO project dealing with the treatment of heavy oil fractions will solve two problems: (1) the need to comply with environmental standards valid after 1998 and (2) an increase in the production of fuels. Eight new plants will be installed to treat heavy oil fractions, thus eliminating the high sulphur content of its products and producing more lead-free gasoline without a higher demand for raw materials. Environmental benefits will mainly take the form of reduced SO_2 emissions and are expected to be quite significant.

over time, however, and therefore it is reasonable to expect that waste producers will eventually change their behaviour to reduce the amount of waste produced in order to reduce their costs. In addition, the penalties which are levied in cases of violation of laws reduce the profit of both landfill operators and generators of waste.

REFERENCES

Ahlander, Ann-Marie Satre (1994), *Environmental Problems in the Shortage Economy: The Legacy of Soviet Environmental Policy,* Aldershot, UK: Edward Elgar.

CowiConsult (1995), 'Case Study of Environmental Expenditure and Investment in Six Selected CEE Countries', Draft Final Slovak Country Report, Prepared for the Organization for Economic Cooperation and Development, June.

Government of Czechoslovakia, Act of the Czechoslovak Government No. 35/1979 on payments in water management as amended by the Act of the Czechoslovak Government No. 91/1988 (complete version No. 2/1989).

Klarer, Jurg (ed.) (1994), *Use of Economic Instruments in Environmental Policy in Central and Eastern Europe,* Budapest: Regional Environment Center, December.

Ministry of Environment of the Slovak Republic (1995), *Environment in Slovakia,* September.

National Council of the Slovak Republic (1992), Parliamentary Law on Charges for Air Pollution No. 311.

National Council of the Slovak Republic (1993), 'Strategy, Principles and Priorities of the State Government Environmental Policy', November.

National Council of the Slovak Republic (1990), Act on the State Administration of the Environment No. 595.

National Council of the Slovak Republic, Parliamentary Law No. 309

1991a on Air Protection as amended by the Law 218/1992; Act of the Slovak Government No. 31

1975 on Penalties for Violation of Obligations for Water Management; Law No. 238

1991b on Waste as amended by Parliamentary Law No. 255/1993; Parliamentary Law No. 287

1994 on Nature and Landscape Protection.

National Council of the Slovak Republic (1967), Act on Administration Procedure No. 71.

World Bank (1995), *World Development Report 1995,* Washington, DC: World Bank.

APPENDIX 10A

Table 10A.1   Summary of charge policies in the Slovak Republic

Subject to payment	Receiver	Source of regulation
Charge: withdrawal of surface water	State Budget	Act No. 138/73, No. 2/89 Ministry of Finance Order
Charge: withdrawal of groundwater	State Water Fund	Act No. 138/73 No. 2/89
Charge: discharge of wastewaters	State Environmental Fund	Act No. 138/73, No. 2/89
Charge: withdrawal from public supply discharge into sewerage	Water Works Enterprises	Act No. 138/73, No. 154/78, No. 15/89, MoF Order 1/93
Charge: rent for hydrostations	River Basin Enterprises	Guideline MoSoil No. 12/63
Penalties: discharge into surface or groundwater without or in breach of a permit, illicit water use, illicit water discharge, illicit disposal of harmful materials, damage of public systems	State Environmental Fund	Act No. 131/73, No. 31/75,
Charge: from large and medium sources		Act No. 309/91
–basic	State Environmental Fund	
–additional	State Environmental Fund	
Charge: from small sources	Municipal Budget	

Table 10A.1　(Continued)

Subject to payment	Receiver	Source of regulation
Penalties: noncompliance with permit, manufacture, export, or import of materials harmful to the air	State Environmental Fund	Act No. 309/91
Noncompliance with permit conditions for small sources	Municipal Budget	
Charge: for solid waste disposal		Act No. 309/92
–basic in technically appropriate sites	Municipal Budget	
–additional in technically inappropriate sites	State Environmental Fund	
Penalties: noncompliance with permits illicit import, export or treatment	State Environmental Fund	Act No. 309/92 Act No. 238/91 Act No. 494/91
Charge: exploitation of raw materials, compensation for extracting area	State Budget	Act No. 491/91, No. 155/94, No. 497/91
Compensation for damage in agriculture or forestry	Agricultural or Forest Enterprises	Order No. 40/63

11. Implementation of Pollution Charges and Fines in Bulgaria[1]

Nikola Matev and Nino I. Nivov

1 INTRODUCTION

The economy in Bulgaria is undergoing a major transition from a command economy to a market-oriented one, and the scope and timing of environmental improvements are closely linked to the success of that transition. In the past the Bulgarian economy was insulated from changes which had a major impact on OECD countries in the 1970s and 1980s: oil price shocks, steep rises in energy prices, restructuring of inefficient heavy industry and shifts towards the service sector.

In the past, Bulgarian industry made little effort to save energy or raw materials. Systematic underpricing led to a substantially higher usage per unit of output than in market economies, resulting in high levels of pollution. This inefficiency was exacerbated by the emphasis on heavy industry and reliance on indigenous fuels.[2] Prevailing technologies, often out of date and inappropriate (for example, those originating in the 1950s and 1960s), promoted an excessive use of natural resources and produced high volumes of waste. Price distortions and subsidies created a bias against investments in pollution abatement.

1.1 Overview of the Structure of Pollution Fines

The system of pollution fines, introduced in 1979 and updated in 1993 and 1995, is considered a key element of Bulgarian environmental policy. The main goals of the pollution fine system are to create incentives for pollut-

---

1   This chapter is limited to air and water pollution charges. The findings and recommendations in the chapter represent the authors' views and not necessarily those of the Ministry of Environment.
2   Particularly low-quality lignite coal.

ers to reduce the pollution load and to raise desired amounts of revenues for the system of environmental protection funds.

In 1979, a system of pollution fines was instituted that applied to air pollution above emission standards, and above ambient standards in the case of water. In 1993, a regulation of the Council of Ministers was passed which updated the system of fines. These revised fine rates, which were actually introduced in 1995, were significantly above previous levels, but did not provide for inflationary adjustment (Council of Ministers, 1995).

Pollution fine rates are linked neither to environmental damages nor to pollution-abatement costs. They are defined based on toxicities of pollutants and on the estimated ability of enterprises to pay them. Pollution fine rates are defined for all pollutants for which there are ambient and emission standards, but in practice fines are applied only to a limited number of pollutants. Air pollutants, such as dust, sulphur dioxide, nitrogen oxides, as well as water pollutants, such as organic materials, suspended solids and heavy metals, are truly regulated by pollution fines. Others are not really regulated.

The fine rates in leva per kilogram (Lv/kg) for selected polluting substances are given in the appendix ($1.00 = Lv 70). The fines are imposed on a monthly basis and are charged until pollution is decreased to permitted levels.

1.2 Environmental Standards

Ambient air quality standards (AAQS) were established by the Ministry of Health in 1969, and were revised in 1984, 1992 and 1994. A partial list of standards is provided in the appendix to this chapter. The standards specify 30-minute and 24-hour average maximum allowable concentrations (MACs) for a great many pollutants (some 180). Some pollutants also have annual standards.

Emission standards were first established in 1978 and were subsequently revised in 1986. They are calculated using actual source data and are based on AAQSs, stack heights, gas flows and temperatures. The formulas that actually set the standards are part of simplified dispersion algorithms. In 1991, new emission standards were introduced for different industrial processes. The standards are differentiated based on the mass flow and the type of polluting substances, classified into four classes according to their toxicities. An example of emission standards for power generation stations is given in Table 11.1.

Ambient water quality standards for three classes of surface waters are defined based on a set of indicators established in 1976 and revised in 1986. Effluent standards are not specified. The first class represents

Table 11.1 Emission standards for power generation stations (more than 50 MW) (mg/m)

Fuel	Plants started before 1992				New plants			
	Dust	SO_x	NO_x	CO	Dust	SO_x	NO_x	CO
Native coal	200	3,500	1,000	250	100	650	600	250
Imported coal	150	2,000	1,300	250	80	650	600	250
Liquid fuel	50	1,700	700	170	50	650	450	170
Gaseous fuel	10	–	500	100	10	–	300	100

potable drinking water, the second is acceptable for recreational use and fish farming, and the third is for irrigation and industrial use. These standards are also given in the appendix.

In many cases the general approach applied requires that all discharges to surface water meet existing ambient standards of receiving waters. It has often been the case, however, that the use of ambient standards as directly enforceable limitations on discharges resulted in unrealistic control requirements, which did not take into account water treatment costs and technologies. As a result, enforcement has been largely ineffective.

2 IMPLEMENTATION EXPERIENCE

Pollution permits exist only for wastewater discharges and are issued by Regional Environmental Inspectorates (REI) of the Ministry of Environment with no expiration date. As was mentioned above, effluent standards do not exist and the general approach applied in permits requires that all discharges to surface waters meet ambient water quality standards for the receiving waters. In practice, this means that ambient standards are used directly as effluent limitations and as a basis for calculating the total amount of pollution fines.

Pollution permits for air emissions do not exist. There are emission standards at the national level for different industrial processes and all enterprises must comply with those standards. The REI is primarily responsible for imposing and collecting fines. There are 16 REIs in the country, each consisting of several divisions dealing with air, water and soil pollution problems, as well as nature conservation. Each REI has its own laboratory and is responsible for both ambient environment and emission compliance monitoring. The staffs of REIs vary from 25 to 55 people.

The main function of the REI is to enforce environmental legislation, to exercise control over the operation of pollution control facilities and to penalize those who violate environmental laws and regulations. The number of enterprises and facilities handled by one inspector in different REIs may vary substantially depending on the size of the REI and the composition of the industrial sector in the region. On average, one inspector is responsible for 40–50 enterprises. More than 50 per cent of an inspector's working time is typically spent on site inspections. If pollution above permissible levels is suspected, the inspector (or chemist from the laboratory of the REI), in the presence of a representative of the enterprise, takes samples for analysis.

2.1 Method for Calculating Fines

Based on the results of the analyses and measurements, inspectors use the following formula to calculate the monthly amount of the fine:

$$A = \sum_{i=1}^{n} Q(K_{fi} - K_{ni}) \times T \times C_i \times 3.6 \times 10^{-3}$$

where:

A the total amount of the fine in leva per month;
Q the flow rate of the wastewater (in l/s) or air pollutant emissions (in m³/s);
K_{fi} the concentration of the i-th polluting substance in wastewater in mg/l; in the case of air emissions, k_{fi} is the concentration of the i-th polluting substance in mg/m³;
K_{ni} the effluent limit or emission standard for the i-th polluting substance in mg/l or mg/m³;
T the period of time over which the discharge is taking place (in hours per month);
C the unit size of the fine for the i-th pollutant in Lv/kg (see appendix for partial list);
i $(1, 2, 3, \ldots, n)$ kinds of polluting substances.

If the calculated amount of the monthly fine for an enterprise does not exceed Lv 50,000, the director of the REI has the power to issue an order for the imposition of the fine. If the calculated amount is more than Lv 50,000, the REI sends the proposed fine to the Ministry of Environment and only the Minister has the right to issue the order. In both cases, the enterprise is notified of the results after receiving the order. In accordance with the provisions of the Environmental Protection Law, the total monthly fine that is allowed to be imposed on a single enterprise is not more than Lv 30 million (approximately $428,000).

The fines are paid out of enterprise profits and are collected monthly. The fines therefore cannot be considered an expense for tax purposes. If pollution is decreased to the permitted level or reduced to some extent, the enterprise notifies the REI by an official letter which describes the new analysis and results. The fine is then cancelled or reduced.

Under the Environmental Protection Law, the Minister of Environment has the right to shut down enterprises if pollution significantly threatens human health or the environment. The threat to shut down factories has been effective in some cases, and as a result pollution-abatement measures have been taken by some polluters. Some strong measures have been taken recently, including slowing production at the Zaharni Zavodi Sugar Industry Complex (Gorna Oriahovitza) to 50 per cent of its capacity. The use of copper ore with high arsenic content was stopped at a copper smelter in Pirdop, and uranium mines that used sulphuric acid injection to extract uranium ore were closed. At the same time it is recognized, however, that the threat of plant closure and the slowing of production are blunt instruments and lack credibility in cases where they would result in significant unemployment and financial losses.

2.2 Revenues from Fines and Their Use

According to the Environmental Protection Act of 1991 (amended in 1992) all revenues from pollution fines are deposited into National and Municipal Environmental Protection Funds. Seventy per cent of the revenues from fines go to the National Environmental Fund and 30 per cent to the fund in the violator's municipality.

National and Municipal Environmental Funds are managed by elected boards, with transparent operating procedures for the allocation of expenditures. The Managing Board of the National Environmental Fund is chaired by the Minister of the Environment and includes deputy ministers and other high-level officials from relevant ministries and institutions.

The main objective of the National Environmental Fund is to support the implementation of the national environmental policy by providing financial assistance (grants or interest-free loans) to municipalities, enterprises and research institutions.

Table 11.2   Annual growth of revenues from pollution fines

Charges/fines for	1989–90	1990–91	1991–92	1992–93
Water pollution (%)	−15.5	−30.5	59.4	126.4
Air pollution (%)	44.6	−50.5	407.5	65.0
Soil contamination (%)	−21.5	11.3	259.0	−28.6
Total	−10.5	−27.7	146.6	58.8
Inflation rate (%)	–	339	91	61

Source: Ministry of Environment.

3   EVALUATING THE EFFECTIVENESS OF THE SYSTEM OF FINES

The fines are ineffective in deterring polluters in Bulgaria. Perhaps most important, they are not related to the environmental damage or the costs of pollution abatement. Assessments of damages to health, agriculture and recreation are neither available nor considered. Second, levels of fines, although significantly increased in 1993 and 1995, are well below estimated pollution-abatement costs and are unable to induce pollution-control actions. Third, fine rates do not keep up with inflation and, as shown in Table 11.2, in real terms the revenues from fines are below 1989 levels.

Enterprises in general know the costs of reducing their emissions, but a main problem in the present economic situation is the lack of finance for environmental investments which inhibits the effectiveness of the pollution fine system. Usually enterprises seek financing from the National Environmental Fund, but its ability to provide financial assistance is very limited.

4   MAIN PROBLEMS THAT NEED TO BE SOLVED TO IMPROVE IMPLEMENTATION AND EFFECTIVENESS OF THE POLLUTION CHARGE SYSTEM

A fundamental problem is that there is no true system of pollution permits. Permits apply only to wastewaters using ambient standards as directly enforceable limitations and there are no plant-specific emission limits. It is therefore very important that an updated system of pollution permits be introduced. A permit should have a set duration (for example, five years) and specify all emission points within the enterprise, set interim and final emission limits with compliance schedules, and establish self-monitoring and reporting requirements. For already-existing facilities, plant-specific requirements should be established following the completion of environmental audits and a series of negotiations carried out at the regional level.

The permit programme should be phased in over a number of years, with the largest sources receiving permits first. For new facilities, a programme of preconstruction design reviews should be implemented immediately. Environmental impact assessments (EIAs), carried out at an early stage of the project cycle, should estimate the effect the new source is expected to have on ambient air, water and soil quality. Following these assessments, design reviews would assess whether the facility as designed complies with pollution-control requirements. Satisfactory results of a design review would allow the applicant to obtain environmental permits. Permits should be renewed no less frequently than every five years.

The levels of fines are low, are not linked to the environmental damages and are not able to change the behaviour of polluters. Compliance with permits should be monitored and fines for pollution above permitted levels should be raised to levels that provide strong incentives for enterprises to adhere strictly to permitted levels (that is, fines should be truly punitive and reflect the seriousness of the violation).

Another problem is that fines do not create incentives for pollution abatement below permitted emissions levels. Introduction of pollution charges was made possible by the adoption of the New Environmental Protection Law in 1991 (amended in 1992), but charge rates have not yet been developed or introduced. Primary targets should be common pollutants which have multiple sources within air control regions and river basins. These include air pollutants such as dust, sulphur dioxide, nitrogen oxides, as well as water pollutants such as organic materials, suspended solids and some heavy metals.

Charges should be considered normal production costs of enterprises and could come from a variety of sources. Targets might include fees for issuing permits, renewal fees which vary based on sizes of sources, emission charges for major sources based on reported emissions or on annual permitted emissions, and fees for inspections to cover the costs of monitoring and enforcement.

A clear distinction must be drawn between (a) fees for emission permits or pollution charges whose purpose is to cover the administrative and monitoring costs and (b) pollution charges which are linked to the amount of damage caused by emissions or which attempt to change enterprises' behaviour. Reliance on cost-based fees to finance regulatory expenditures are common in Western Europe and should be regarded as a minimum requirement to ensure that environmental authorities have sufficient funds to fulfil their responsibilities for monitoring and enforcement. Pollution charges would then generate additional government revenues that could be used to finance environmental expenditures via environmental funds.

The Ministry of Environment has recently taken steps to develop and introduce pollution charges into the enforcement mechanism. For example, we intend to introduce a system of water pollution charges. It is expected that charge rates will be set to be consistent with consumers' abilities to pay and the system will be designed to be easily implementable. Rates will be calculated per cubic metre of fresh water used and differentiated by the quality of wastewater discharged. It is hoped that such a system of water pollution charges will encourage water conservation, recycling and wastewater quality improvements. Revenues from these new charges will be accumulated in the Environmental Fund and earmarked for completion and new construction of municipal wastewater treatment plants. These charges are expected to complement the effects of new increases in water tariffs by water companies throughout the country, and also provide additional resources for investments in wastewater treatment.

REFERENCE

Council of Ministers, Republic of Bulgaria (1995), 'Regulation on the Calculation and Imposition of Fines Applied to Pollution of the Environment Above Permissible Levels', *State Gazette*, No. 15, amended No. 101.

APPENDIX IIA

Table 11A.1   Partial list of Bulgarian ambient air quality standards*

Compound	Maximum allowable concentration (mg/m^3)		
	30 minute	24 hour	Average annual
NO$_2$	0.20	0.10	0.05
Benzene	1.50	0.10	
3,4-benzopyrene	–	0.10	
Benzoil chloride	0.06	0.03	
CO	60.00	10.00	
Ethylene	3.00	3.00	
CdO (calculated as Cd)	–	0.00001	0.00001
MgO	0.40	0.05	
Methanol	1.0	0.5	
Phenol	0.01	0.01	
Zinc	–	0.05	
Ni-oxide	–	0.001	
Lead compounds	–	0.001	0.001
O$_3$	0.160	0.10	
Dust – non-toxic	0.50	0.25	0.15
H$_2$SO$_4$ (calculated by molecule)	0.30	0.30	
SO$_2$	0.5	0.15	0.05

Note: *A full list of standards is available from the authors or editors.

Table 11A.2   Partial list of Bulgarian ambient water quality standards*

	Maximum allowable concentration (mg/l)
Physical and chemical parameters	
Temperature (C°)	not in excess of 3° of the average temperature of the season
Dissolved oxygen	< 40
Dissolved solids	< 1,000
Suspended solids	< 50
Hydrogen sulphide (free)	not allowed
Nitrogen (ammonia)	< 2.0
Phosphates	< 2.0
Organic pollution	
Suspended substances	< 15
Bichromate COD	< 70
BOD_5	< 15
Organic nitrogen	< 5
Inorganic substances of industrial origin	
Mercury	< 0.001
Cadmium	< 0.01
Lead	< 0.05
Arsenic	< 0.05
Copper	< 0.1
Chrome (3-valent)	< 0.5
Nickel	< 0.2
Zinc	< 5
Fluorides	< 1.5
Organic substances of industrial origin	
Phenols (volatile)	< 0.05
Petroleum products	< 0.3
Pyridine	< 0.2
Benzene	< 0.5
Formaldehyde	< 0.5

Note: *A full list of standards is available from the authors or editors.

Table 11A.3   Partial list of pollution fines

	Unit size of the fine in LV/KG
Water pollutants	
1. Suspended solids	3.00
2. Oxidizability (permanganate), BOD, COD*	8.00
3. Nitrogen from ammonium	16.00
4. Petrol products	112.00
5. Phenols	640.00
6. Hydrogen sulphide	480.00
7. Mercury	48,000.00
8. Cadmium	4,800.00
9. Lead	960.00
10. Arsenic	240.00
11. Copper	96.00
12. Chromium (threevalent)	96.00
13. Nickel	240.00
14. Zinc	9.60
15. Formaldehyde	64.00
Air pollutants	
1. Nitrogen dioxide	3.60
2. Ammonia	1.60
3. Cadmium	400.00
4. Copper	40.00
5. Lead	1,320.00
6. Dust (non-toxic)	0.80
7. Sulphur dioxide	1.20
8. Hydrogen sulphide	36.00
9. Soot	2.80
10. Arsenic	132.00
11. Carbon sulphide	36.00
12. Hydrochloric acid	1.60
13. Sulphuric acid	148.00
14. Phenols	30.00
15. Chlorine	3.20
16. Zinc	8.00

Notes:
Exchange rate US$1 = Lv 70.
*The fine by point 2 is imposed for only one of the three indicators, specified by the supervising body depending on the wastewater kind.

12. The Road to Creating an Integrated Pollution Charge and Permitting System in Romania

Clifford F. Zinnes[1]

1 INTRODUCTION

Romania is a country of 23 million people with abundant natural and environmental resources. Water resources include the Danube River and twelve tributary basins, as well as part of the Black Sea. Forest resources include 6.3 billion hectares of forest, covering 26 per cent of the country's surface. In addition to deposits of ferrous and non-ferrous metals, the country has reserves of oil, natural gas and coal. Romania is also home to the 650,000-hectare Danube Delta, comprising the largest wetland in Europe.

This resource abundance, however, has had its costs. The country developed minerals, petrochemical and metals-processing industries that are highly polluting, leading to economic, health and ecological impacts on an enormous scale. Such impacts have also stymied the development of activities with a potential future, such as tourism and fisheries.[2] While these costs are clear, given four years of negative GNP growth during the period 1992–95, there is understandable trepidation about pursuing too quick or rigorous a programme of environmental protection.

With the adoption of a new Framework Environmental Law No. 137/1995, passed on 31 December 1995 after three years of wrangling, and with an almost approved water law, Romania is at a crossroads in its efforts to create a legal basis for its transition to an environmentally sustainable market economy. These new laws allow for a number of policy instruments

1 The author is particularly indebted to Dr Mihaela Popovici for her input into Sections 2.3 and 3.2, as well as for comments on other parts of this chapter.
2 A summary of the worst of these impacts by key economic sector and by health, ecological and economic effect is provided in Manea and Zinnes (1994).

to be used, including environmental permits, user charges (including for pollution), subsidies, legal liability and other economic incentives.

This chapter provides an overview and analysis of the experience with implementing pollution permits and related payments, with a focus on air and water.[3] Some of these payments are direct user charges based on the quantity of a resource used (for example, m^3 water consumed), some are direct charges on pollution (for example, tons of SO_2 or BOD emitted), and some are penalties for violating specific conditions in permits that are not directly based on a quantity unit. The chapter also attempts to develop a number of broad themes concerning the passage and implementation of environmental legislation in a transition economy.

2   AN INTRODUCTION TO PERMITS AND CHARGES IN ROMANIA[4]

2.1   Introduction

Romania has a centralized administrative system for managing the environment that is the responsibility of the Ministry of Water, Forests and Environmental Protection (MAPPM). Established in 1990, this ministry is in charge of developing, implementing, monitoring and enforcing legislation related to the protection of air, water, forests and soils. The ministry comprises three main departments: water, forests and environmental protection. The ministry also includes a commission for nuclear safety and a large research and engineering institute called the Institute for Environmental Research and Engineering (ICIM). Local governments are responsible for land-use planning, water supply, wastewater treatment and municipal waste collection and disposal.[5]

For implementation and enforcement of environmental laws and regulations, responsibilities are delegated to the 41 branch agencies and the Danube Delta Biosphere Reserve Administration under MAPPM's Environmental Protection Department. There is one branch agency per county (called *judeti*). These local Environmental Protection Agencies (EPAs) are generally responsible for issuing permits, monitoring and

3   See Regional Environment Center (1994) for a description of instruments related to waste.

4   The permitting system in Romania has just been completely revised. This section describes the system up to the present; the final section of this chapter outlines how the system should differ in the future.

5   The Ministry of Agriculture also has some responsibilities for soil protection, and can require that compensation be paid to farmers for damages to agricultural lands.

charge collections. The Environmental Protection Department of MAPPM is responsible for regulatory and legislative drafting, as well as for more complex construction permit decisions.

The Water Department within MAPPM consists of an inspectorate and two directorates.[6] The Water Department also supervises *Apele Române* (AR), a public utility with branches in each of the country's 12 river basins. AR is responsible for the management of 70,000 kilometres of rivers and 150 multipurpose lakes and dikes. AR supplies 95 per cent of the raw water to municipalities, industry and agriculture.

2.2   Air Permits

Ambient air standards are set by the Ministry of Health, with the right to fine for infringing these standards set by Law 9/1973 and MAPPM Order 127/1994. Environmental standards are typically drafted by research institutes and approved by the affected ministries, usually with MAPPM as the sponsoring ministry. Sample air standards are given in Table 12.1. MAPPM attempts to achieve these standards through a system of 'environmental authorizations' covering all media.[7] All requests for authorizations are submitted to local EPAs, where they are either reviewed or passed on to the Environmental Department at MAPPM. The system of environmental authorizations has three possible components. These are environmental agreements, environmental permits and so-called 'plans of measure'.[8]

An environmental agreement sets the conditions for carrying out a *new* project or activity and for modifying an existing facility. Documentation necessary for obtaining the agreement for air pollution emissions includes information on technological conditions of the enterprise, data on the existing ambient environment, projected emissions (including noise), proposed abatement technologies, and plans for solving any environmental problems related to transport. In many cases an environmental impact assessment (EIA) is also required to obtain an environmental agreement from local EPAs or MAPPM. Whether jurisdiction rests with the EPAs or the Environmental Department typically depends on the size and complexity of the activity. About 250 agreements were issued annually over the period 1993–95, mostly to commercial companies with

6   One for strategic planning, drafting laws, issuing regulations and interministerial coordination, and one for hydrological and meteorological assessment and flood control.

7   Thirty-four laws and regulations form the legal basis for issuing authorizations. They are listed in MAPPM (1995) and many are described in Regional Environment Center (1994).

8   This is a technical term used by MAPPM. It implies the development of a plan by which a firm agrees to improve its environmental performance within a specified time frame.

Table 12.1   Maximum permissible concentrations of selected air pollutants

Pollutant	Units	Sampling duration		
		30 minutes	24 hours	Monthly
Ammonia	mg/m^3	n.a.	0.1000	n.a.
Sulphur dioxide	mg/m^3	0.7500	0.2500	n.a.
Ozone	mg/m^3	0.1000	0.0300	n.a.
Carbon monoxide	mg/m^3	6.0000	2.0000	n.a.
Nitrogen dioxide	mg/m^3	0.3000	0.1000	n.a.
Sulphuric acid	mg/m^3	n.a.	0.0120	n.a.
Lead compounds	mg/m^3	n.a.	0.0007	n.a.
Cadmium	g/m^3	n.a.	0.0200	n.a.
Deposited powder	g/m^3	n.a.	n.a.	17.000
Suspended powder	mg/m^3	0.5000	0.1500	n.a.
Hydrogen sulphide	mg/m^3	n.a.	0.0080	n.a.
Fluorine compounds	mg/m^3	n.a.	0.0050	n.a.

Note: n.a. = not applicable.
Source: National Commission for Statistics (1996) pp. 54–61.

the state as the major shareholder. This figure is set to rise dramatically once the Framework Environmental Law goes into effect, increasing by an unmanageable hundredfold if the law is scrupulously followed.

Once the environmental agreement has been procured and the project's construction has been completed, an environmental permit is sought. This establishes the operational conditions and technical specifications under which existing and new activities are allowed to function. The environmental permit is given only after basic operation permits have been obtained from other licensing agencies. Emissions standards for stationary sources are set in MAPPM Order 462/1993. All permits specify that enterprises are responsible for monitoring emissions and for reporting to EPAs.

Although EPAs have the right to amend, suspend and revoke environmental authorizations, they typically prefer to be more collaborative in reaching compliance. If an enterprise is unable to meet the discharge levels specified in its permit, a plan of measure is developed containing the steps that must be undertaken within a specified time frame. If these steps are not taken, the plant must be shut down within two years. The maximum allowed term for bringing an enterprise into compliance is seven years (beginning 1 July 1993).

2.3 Water Consumption and Effluent Permits

Water resources in Romania are administered according to the principles of integrated water management. Policies to promote sustainability of the resource therefore try specifically to incorporate links between water quality and water quantity. This linkage is potentially important, because excessive abstractions lower underground- and surface-water levels and increase contaminant concentrations. Thus, excessive abstraction can have just as deleterious environmental effects as effluent discharges. The permitting process, the heart of the regulatory system, incorporates this duality by issuing permits and assessing charges and fines both for water consumption and for effluent discharges.

Apele Române (AR), with its 12 river basin branch offices, issues permits based on guidance from the MAPPM Water Department and the national water-management strategy. These permits specify the amount of water used or consumed, as well as the quantity and quality of effluents. Permits do not focus on discharges if wastewater is emitted into a municipal sewage system. Permits are generally valid for a year, but some facilities receive shorter durations. For example, nuclear power plants receive three-month permits.

Effluent standards for discharges are set in Decree 414/1979 and apply to any potential source of pollution involved in the generation, transport, storage, treatment or disposal of wastewater. Ambient water quality standards were approved in Decree 4706/1988. Examples of these standards are shown in Table 12.2.

The Water Department of MAPPM and AR both have important roles in enforcement. For example, both the MAPPM Water Department and local branches of AR can take legal action against noncomplying facilities and levy fines and other sanctions against violators. The MAPPM Water Department has the authority to inspect and is responsible for determining if facilities are in compliance with water management requirements. Although enterprises are responsible for monitoring and reporting their emissions, data are also collected by AR from a series of sampling checkpoints along water bodies.

Water management standards include effluent standards that limit the amount or rate of discharges. These standards provide some flexibility, because they allow facilities to choose which technologies should be used to meet requirements, thus reducing compliance costs. Compliance is monitored by periodic sampling and by inspecting the records of facilities.

Table 12.2 Ambient standards for rivers and for discharges

Indicator	River quality class*			Level of dilution (mg/dm^3)**		
	I	I	III	I	50	100
Suspended materials	n.a.	n.a.	n.a.	25	100	200
BODs	5	7	12	15	60	100
Hydrogen sulphide (H_2S)	n.a.	n.a.	0.1	0.1	1	2
Phenols	0.001	0.02	0.05	0.02	0.3	0.6
Lead (Pb)	0.05	0.05	0.05	0.2	0.2	0.2
Cyanide CN$^-$	0.01	0.01	0.01	0.1	1	2
Total ionic iron (Fe^{2+})	0.3	1	1	2	5	8
Nitrogen dioxides (NO_2^-)	10	30	n.a.	n.a.	n.a.	n.a.

Notes:
* Romanian rivers are divided into three quality classes, each with ambient standards and permitted uses.
**The dilution represents the ratio between the minimum annual mean monthly flow providing for 95 per cent of the emissary and the total water discharge flow.
Source: STAS 4706/1988 and Decree 414/1979.

There are also, however, several best-practice norms that can be used to regulate facilities. These practices may include requirements that approved equipment be located in a specific place and that it operate properly. Thus, although water authorizations are based primarily on performance standards, AR may still require that certain practices be used, because they are easily implementable, or they will improve a firm's ability to monitor its discharges, or they address certain local peculiarities.

3 DIRECT CHARGES ON POLLUTION QUANTITIES

3.1 Air Pollution Charges

There are no true air pollution charges in Romania. There are penalties for exceeding permitted emission levels, but these are implemented on a case-by-case basis and are not linked to pollution quantities through, for example, some type of algorithm. Penalty levels depend on the polluter's legal status and can vary between 250,000 and 500,000 lei (US$100–200 at 1995 exchange rates) for physical persons and between 500,000 and 1,000,000 (US$200–400) for legal persons.

Penalties are collected by local EPAs, who seem to have a great deal of discretion in setting penalties. Given local EPAs' limited staff and larger concerns, collecting penalties from physical persons is simply not done. For enterprises, penalties for air emissions provide no deterrent effect; their purpose is perhaps only one of education. It is possible that some system of air pollution charges, based directly on emissions, will be implemented in the future under the Framework Environmental Law. However, with the exception of fees for permits, there is little explicit in the new law as to how such charges would be implemented.

3.2 Water Charges

Water charges exist in Romania, both for direct consumption or use and for discharges. Their aims are to encourage sustainable resource use and to generate revenues to finance water supply and sewage treatment and disposal. Water charges were introduced at the start of 1991 and rates are indexed quarterly. There are separate national prices for each category and user of raw water, with industry paying more than agriculture, and agriculture paying more than households. There is a separate price for water used by thermal power stations (heating and hot water).

Water withdrawn from the Danube, inner rivers, lakes and groundwater all have separate prices. With little relationship to location-specific or temporal characteristics, however, this system does not accurately signal differences in water scarcity to economic agents. As a result, water-intensive activities may be undertaken in high-cost zones, though in principle AR could reject a construction permit request for such an activity. Water rates per cubic metre in 1994 for industrial users were 2,593 lei (US$1.30) from rivers, 315 lei (US$0.16) from the Danube, and 3,194 lei ($US1.60) from underground sources.

The charge for raw water to municipalities for drinking water averages 14 lei (US$0.007) per cubic metre. The municipalities, in turn, charge households between 10 and 100 times that amount. Although households often find water charges high, they probably do not encourage conserva-

9   See DeShazo et al. (1996) for an analysis of municipal water pricing and household demand in Romania.

tion, because water bills are computed based purely on household size, and consumption is not metered.[9] Penalties of two to six times normal rates can be levied for abstractions above permitted limits, with the multiplier depending on the amount of the infraction and whether it occurred during a period of restricted consumption.

In addition to pricing water, there are also charges for effluent discharges into water bodies. This system depends on both the volume of wastewater discharged and the difference between actual and permitted concentrations. For the *i*-th pollutant, the formula used is:

$$P_i = (C_i - C_i^*) \, V \, R_i$$

where P_i is the total penalty assessed on the *i*-th pollutant, C_i and C_i^* are, respectively, the actual and permitted concentrations of the *i*-th pollutant, V is the annual volume of wastewater discharged, and R_i is the rate for discharging the *i*-th pollutant. Up to the allowable concentration the payments are zero, but for repeat offenders penalties are doubled each year until concentration standards are met.

Penalties are levied on twenty substances divided into two general categories. The first group (containing substances such as BODs, chlorine, cyanide, CODs, sulphates, cadmium, nitrates, detergents, ammonia and arsenic) is for those for which allowable levels are established to meet concentration standards. The second group (containing substances such as mercury, persistent pesticides, radioactive residues and carcinogens) is made up of substances for which no discharges are permitted and C_i^* is zero. Rates are lower for the first group of pollutants than for the second group. For example, 6.58 lei/kg is charged for suspended matter, while the rate is 6,595.33 lei/kg for cadmium and phenols, 79,144 lei/kg for mercury and 329,770 lei/kg for cancerous substances.

All revenues collected by AR are returned to the national treasury. Each year the central office of AR receives a budget allocation, which it distributes after covering central office costs. AR has been able to cover only approximately 75 per cent of its costs, however, with the rest coming from additional treasury allocations. Indeed, the greatest weakness of the current system is the inadequate cost recovery for network maintenance and expansion. One reason is that cost accounting is not based on replacement value, a very serious problem in a country running a high inflation rate. A second reason is that long-term financial planning is not possible because government financial controls are set on a short-term basis and carryover is not allowed. To address this financing problem partially, a water fund was created and capitalized by earmarking five per cent of revenues from water charges and 100 per cent of penalties.

Unfortunately, enterprise arrears, non-payment and low penalty rates have meant that funding is still inadequate.[10]

4   AN ANALYSIS OF THE CURRENT AIR QUALITY MANAGEMENT SYSTEM

4.1   Implementation Experience with Air Pollution Permits and Emission Controls

The problem with the permitting system is not a lack of regulations and laws. The problem is that upwards of 90 per cent of Romanian polluters are operating without valid permits. The basic truth about the system described above is that it is simply not enforced. As illustrated in Figure 12.1, it is tempting to blame unrealistic standards being imposed on industry; with few exceptions Romania's standards are stricter than internationally accepted norms.

In addition to enforcement-related problems, the generally poor response of enterprises to environmental regulations is related to a lack of adequate financing[11] and a general reluctance to invest in end-of-the-pipe solutions for plant and equipment which could then be replaced during enterprise restructuring after privatization. Until recently, enterprises' principal source of funding for environmental investment was the state budget. Such funds were allocated through the overseeing ministry – typically the Ministry of Industries or the Ministry of Environment – but these funds are drying up fast. For example, the national environmental action plan envisions future state budget funding for only 45 per cent of all environmental investments.

Although the non-governmental organization sector is quite large in Romania, there are no explicit mechanisms for public participation in the environmental permitting and enforcement processes. Non-governmental organizations have, for the most part, focused on commenting on draft legislation and on developing a competing national environmental strategy for the government. It should also be noted that enforcement of regulations has been reduced, because of a battle between local EPAs and their ministry (MAPPM) and enterprises and their ministry (historically the Ministry of Industries, but more recently the State Ownership Fund, which holds 70 per cent of the shares of most companies). Without strong

10   It is hoped that the draft water law will be passed, in which the 'polluter pays' and 'beneficiary pays' principles are clearly stated. This draft law also includes plans for a number of water funds.
11   Many of the worst offenders are indeed legally prohibited from further borrowing.

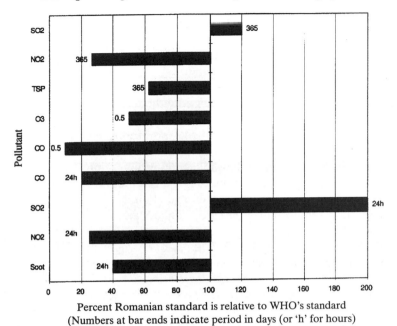

Percent Romanian standard is relative to WHO's standard
(Numbers at bar ends indicate period in days (or 'h' for hours)

Figure 12.1   *Comparison of Romania's ambient air quality standards*
with WHO recommendations

public support, MAPPM, under current transition conditions, usually cannot win such battles.

Local environmental protection agencies are grossly understaffed, underequipped and underpaid for the work they are required to carry out. In 1994 there were 450 local EPA inspectors in the country. According to ICIM, the research institute of MAPPM, these inspectors visited 2,900 facilities on a regular basis – typically two to twelve visits per year – and made a total of 52,100 visits. From the inspections made during these visits, 6,234 violations were registered, implying that 12 per cent of inspections led to violations. This low rate at least suggests that facilities, although inspected, are not fined when in violation of permit requirements.

One dramatic example of enforcement occurred in 1994 when a local EPA forced the closure of ROMPLOMB, a lead smelter in the middle of Baia Mare, a city of 160 thousand people. Although considered a major victory by environmentalists, in 1994 ROMPLOMB was the only closure, and there were relatively few other enforcement actions taken that year.[12] This

12   Only 25 permits were withdrawn in 1994 and twelve permits were suspended. Only six criminal prosecutions were made in 1994.

relatively lax enforcement is perhaps not surprising. Local EPA directors are picked by the governor and the Minister of Environment, both appointed by the governing party of the central government. With unemployment the key concern of the government, if directors pursue their jobs too vigorously they are simply dismissed. If they are too permissive, sometimes (but not too often) local populations take them to task, in which case the director is also sacked. The result is that local EPAs can be difficult places to work. In addition, when they collect and present environmental data they are often chastised if the data imply that emissions are not improving, especially if the government has invested in that region. This situation is not likely to change except to the degree that public participation and, more importantly, public interest increases. Public support was, for example, particularly important in the case of Baia Mare.

MAPPM finds itself in many ways in a position similar to that of local EPAs. With a staff of only 15 individuals, for example, the Environmental Protection Department was given only three months to revise and submit revised permitting and compliance legislation to parliament. It had to do this with no legal support, because MAPPM has no lawyers. Not surprisingly, drafting legislation without lawyers results in poorly drafted laws, but the situation is unlikely to change in the near future, because salaries are too low to attract qualified individuals. Indeed, salaries for department directors are set at $100 per month.

Turning to air pollution penalties, and again using 1994 data, approximately 1 billion lei of penalties (US$625,000) were levied in 1994. With 6,234 violations, this means an average of 168 thousand lei (roughly US$105) per violation. According to the statutes, the minimum and maximum penalties are 500 thousand lei (US$250) and 1 million lei (US$500), respectively. The average fine was therefore even less than the minimum specified in the legislation, and this low level is unlikely to provide much economic incentive to undertake abatement investments. Note that fines, in any case, pass directly to the Ministry of Finance; thus, even the *institutional* incentives to enforce the statutes are not very strong. Nevertheless, it is still worth improving the system to collect these fines even though they are low (perhaps lower than collection costs). This is because experience in other countries shows that it is easier to establish a well-functioning collection system with low rates than with high ones. Once an improved system is in place, rates can then be raised as mandated by the new framework environmental law.

4.2   State-sponsored Investment Subsidies for Air Pollution Control

Because of the inability of environmental authorities to impose penalties or use other economic instruments to achieve acceptable emissions, and

Table 12.3   *Sources of funds for stimulating environmental abatement*
            (billions of 1995 lei)

Source	NEAP*		1994, air abatement	
	Billions of 1995 lei	Percentage	Billions of 1995 lei	Percentage
Own sources	159	17	123	42
External	358	38	0	0
Budget	417	45	168	**58
Total	934	100	290	100

Notes:
* NEAP stands for National Environmental Action Plan and includes all media.
**However, the proposed investment programme implied a level of 61 per cent.
Source: MAPPM (1995) and internal MAPPM sources.

because of enterprises' indifference to operating without valid permits, environmental authorities have focused more on subsidizing pollution control and less on increasing the costs of air emissions. The favoured instrument of the government has been to augment the permit system with environmental abatement investment co-financing to stimulate enterprises to fulfil permit requirements. As shown in Table 12.3, in 1994 the national treasury provided the majority of financing. Although this mechanism is likely to continue, the government's contribution seems set to fall, with most of the burden being shifted on to economic agents themselves. In the country's National Environmental Action Plan, the budget's share of investment will fall to 45 per cent of all investments compared to 58 per cent in 1994.[13]

Given the importance of public investments in environmental protection, and the reliance on investment subsidies, it is perhaps useful to try to identify the government's observed investment preferences and attempt to evaluate whether they correspond with stated government priorities and conventional wisdom regarding the effects of various pollutants on human health. These inferences are made by examining relationships between the amounts spent by *judeti* and the seriousness of the pollution problems in those regions in 1994. To give some indication

13   It remains to be seen whether external sources (primarily international donors) actually pick up a substantial portion as expected. If not, the state's share could increase or the level of investments will have to be scaled back considerably.

of what problems attracted government money, pairwise correlations between various measures of pollution 'seriousness' and aggregate investment expenditures are presented in Table 12.4. Data on two key measures of pollution are used for each of the 41 *judete*. The first measure is the frequency during 1994 that maximum permitted concentrations were exceeded (these measures end in '–%'). The second type of indica-

Table 12.4　Pairwise correlation between judeti-*level investments in air pollution abatement and various* judeti *characteristics*

Variable name	Type	Description	Correlation*
POWDER–%	Frequency	Deposited powder	0.170
AMMONIA–%	Frequency	Ammonia	–0.040
LEAD–%	Frequency	Lead and lead compounds	0.550
LEAD–AV.YR	Average	Lead and lead compounds	0.734
SO_2–%	Frequency	Sulphur dioxide	0.400
SO_2–AV.YR	Average	Sulphur dioxide	0.577
NO_2–%	Frequency	Nitrogen dioxide	0.207
NO_2–AV.YR	Average	Nitrogen dioxide	0.013
DRYINGSTR	Quantity	Area of forest showing signs of extreme drying	0.004
DRYTOT	Quantity	Area of forest showing any signs of drying	0.153
DRY*SO_2	Quantity	Product of DryTot times SO_2–av.yr (indicator of where SO_2 was leading to forest damage)	0.257
LF_IND/POP	Quantity	LF_ind/Population (indicator of industrial activity)	0.283
HOTSPOTS	Quantity	Number of World Bank 'hot spots' in the county	0.094
POPULATION	Quantity	Population	–0.050
LFIND	Quantity	Labour force in industry (extractive, manufacturing and energy)	0.065

Note:　*With state investment in air pollution abatement.
Source: Author's computations based on data from National Commission for Statistics (1996).

tor is the yearly average concentration (in milligrams per cubic metre) (these end in '–av. yr').

Perhaps the major result of the analysis is that investments do tend to be made in the highest priority pollutants. The pairwise correlations presented in Table 12.4 suggest that government money was particularly drawn to areas where lead and sulphur-dioxide emissions were major problems. Correlations were much lower between investment expenditures and nitrogen-dioxide and deposited powder emissions, at least partially reflecting their lower position in the priorities of the government and the international community. The low correlation coefficients for POPULATION and LF_IND, however, suggest that the number of beneficiaries alone is not an extremely important factor. The correlations also suggest that in 1994 investments were much more spread out than just to those areas identified by the World Bank as 'hot spots'.[14]

5   IMPLEMENTATION EXPERIENCE WITH WATER QUALITY MANAGEMENT

5.1   Use and Effectiveness of Water Charges

As pointed out in Section 2.3, integrated water management is based on the notion that both excessive abstraction and contaminant discharge pose threats to the environment. Penalties are for effluent discharges and raw water abstractions *above permitted limits* only. In 1993 there were 1,772 charges assessed, yielding collections of 254 million lei (US$250,000). Penalties levied fell to 1,445 in 1994 leading to receipts of 546 million lei (US$342,000).

Revenues collected, however, were only about one-quarter of the total penalties assessed in 1993 and one-third of those assessed in 1994. Collection is therefore a serious problem even with such low rates. With the hope of increasing its collections of water charges, MAPPM implemented a phase-in programme in 1991 for payments of assessed penalties. It was announced that for 1991, 25 per cent of the assessed penalty must be paid, 50 per cent of what would be assessed in 1992 should be paid, 75 per cent in 1993, and 100 per cent from 1994 onwards, and starting in 1996, fines owed would double each year until individual standards are met.

---

14   The differing values placed on various pollutants are quantified in Zinnes (1996), where it is shown that a one per cent increase in the frequency that nitrogen-dioxide concentrations exceeded the maximum allowable in 1994 was correlated with an increase in investment co-financing of $68,500. This contrasted with a value of only $10,000 for deposited powder emissions.

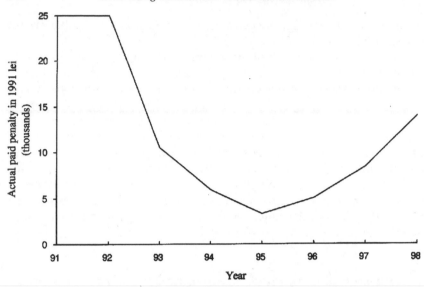

Figure 12.2   Real value of a 100-thousand lei penalty (in 1991 lei) in
various years under the past penalty payment scheme

The goal of this phase-in is to increase the pressure on enterprises over time and allow them to catch up. The schedule was not indexed for changes in prices, however, and inflation averaged 175 per cent per year during the period 1991 to 1996. Thus, exactly the opposite effect was achieved. As shown in Figure 12.2, in real terms the pressure on enterprises actually fell over time up until 1995 and only after the period of forgiveness ended did real penalty rates increase.

5.2 Assessing Effectiveness of Enforcement Activities

Using assessed penalty and collection data at the level of the river basin from 1993 and 1994 it is possible to look at how the twelve ARs have been coping with enterprises that do not pay their water pollution and water abstraction penalties.[15] We therefore construct the following two variables:

15   For 1993 and 1994 respectively, the following basin averages were observed: number of units inspected 485 and 734, units penalized 129 and 145, value of fines assessed in current lei 88 million and 166 million, value of fines collected in current lei 12 million and 61 million. The inflation rate at the producer level in 1994 was 137 per cent. In 1993 two-thirds of the fines by *judeti* were in the US$230 to 600 range and in 1994 two-thirds were between US$1,200 and 3,600.

$$\text{Collection rate in basin } i = \frac{\text{Value of penalties paid in basin } i}{\text{Value of penalties assessed in basin } i} \equiv \text{COLRATE}_i$$

$$\text{Mean real penalty size in basin } i = \frac{\text{Value of penalties assessed in basin } i}{\text{Number of penalties assessed in basin } i} \equiv \text{PENSIZE}_i$$

With this information we examine whether average penalty size has any relationship with collection rates. It was found that correlations between average penalties and collection rates for 1993 and 1994 were −0.27 and −0.37, respectively. These results suggest that because returns to evasion are increasing in the value of penalties, higher penalties tend to reduce collections; smaller penalties therefore appear to be easier to collect. Second, the change (denoted by Δ) in collection rates by river basin from 1993 to 1994 is regressed on the change in the average real penalty size and its square for the period 1993 to 1994. The following results were derived:

$$\Delta\text{COLRATE}_i = \underset{(t=2.00)}{0.217} - \underset{(t=-4.32)}{0.659*\Delta\text{PENSIZE}_i} - \underset{(t=-2.50)}{0.392*(\Delta\text{PENSIZE}_i)^2}, n=10, R^2=0.73$$

Although the model is very simple and the sample size is small, the results strongly suggest that increases in penalty sizes between 1993 and 1994 tended to be associated with lower collection rates. The negative sign on the squared term, however, indicates that the collection rate cannot be increased indefinitely by lowering the size of penalties.[16]

Another question of potential interest is whether inspectors find more infractions when they inspect facilities more frequently. This information may be quite useful, because an answer would provide an indication of the response of polluters to the use of more frequent monitoring to complement the use of fiscal instruments such as penalties. *A priori*, it is indeed unclear whether an inspector would find more problems or less when facilities are inspected more frequently. More inspections are, for example, likely to find more infractions if polluters do not change their behaviour. On the other hand, if enterprises change their behaviour, for example, by more careful monitoring and control of production operations, or if inspections are less thorough due to a limited number of inspectors carrying out more of them, the relationship could be negative; in this case more fre-

16   In Zinnes (1996) this relationship is used to identify the revenue-maximizing average penalty size.

quent inspections would imply that environmental performance would at least appear to improve.

This question was investigated in two ways. First, using these same data we now define a variable for the infraction rate in river basin i. This variable is defined as:

$$\text{PENRATE}_i = \frac{\text{Number of infractions identified in basin } i}{\text{Number of facilities inspected in basin } i}$$

Examining changes in PENRATE between 1993 and 1994 and percentage changes in the number of inspections that occurred over the same period (%ΔNUMINSP) reveals a negative relationship. Second, inspection elasticities of infraction detection were computed for each river basin revealing half having negative and half having positive elasticities. Together these results suggest that while enterprises in general have responded to an increased frequency of inspection by reducing the number of their violations, this has occurred at the price of a lower yield or efficiency of inspections, indicating that either inspectors became more over-extended or that the additional inspections were carried out at sites with better environmental performance. The results also point to a substantial variability of enforcement across the country.[17]

6   LIKELY IMPROVEMENT DUE TO NEW OR PENDING LEGISLATION[18]

Romania has come from a history of excessive regulation where many of the rules were so absurd as to be unenforceable. The result has been a vicious circle of drafting of environmental regulations containing standards, procedures and penalties that everyone knew from the beginning would be ignored; because they were not enforced, enterprises applied no pressure on regulators to make more realistic regulations so the regulators apparently focused on the propaganda value of increasingly strict – and therefore ever more unenforceable – regulations. Moreover, low

---

17   For the first analysis, an OLS regression was run of PENRATE_i^{94} on a constant, PENRATE_i^{93}, and %$\Delta\text{NUMINSP}_i$ and indicated that a ten-percent increase in inspections led to a one-percent decrease in the rate of infractions. For the second analysis, elasticities indicated what the effect of a percentage change in the number of inspections had on the percentage change in the number of infractions discovered, and varied in general from -1 to $+1.3$. Zinnes (1996) discusses in greater detail the reasons for these results.

18   The author has collaborated closely with the Romanian environmental authorities in the drafting of the new legislation and regulation described in this section.

fines and charges often were not collected, creating a belief that enter prises had tacit permission to pollute and there was no reason to try to improve. Can the future be any different?

Looking first at air permitting reform, what has been missing from the permitting system and, in fact, environmental policy as a whole, has been legislation on environmental sustainability and public participation. Without public participation and public interest the environmental permitting system is frequently seen as obstructing economic activity, making it difficult for regulators to do their jobs. Public participation, in turn, was missing for two reasons. First, information was restricted. A perfect example is the lack of public knowledge of the effects of lead on children. Second, with a few exceptions, mechanisms were not in place to integrate the public into the decision-making process. The new framework environmental law addresses this problem by giving the right to any party to force government agencies to enforce environmental permits and their charges.

New regulations mandated by the framework environmental law introduce several permit innovations to assist with implementing the charge and permit system. First, plans of measures were generally not developed with a goal of minimizing the costs of compliance. Plans were therefore more expensive to carry out than necessary and these higher costs deterred their implementation. In the new framework environmental law, integrated permitting is encouraged by establishing temporary technical review teams comprising representatives of all relevant government bodies involved in the environmental permitting process (such as forestry, health, water and soil). By negotiating compliance schedules – which will replace plans of measures under the new framework law – directly with these teams, enterprises will save time and money and the outcome is expected to be more permitting consistency across regulatory bodies.

Second, inappropriate public investments were often made because decisions were made based on limited and often unreliable information. In the new law, permits will require a toxic release inventory even for those emissions not subject to permitting limits. In addition to facilitating the identification of infractions, this is probably the cheapest and most efficient way for local EPAs to gather the information necessary for environmental planning and policy decision making; it also helps to sensitize enterprises regarding their waste streams.

Third, plans of measures, though technically feasible, could often not be carried out due to lack of funds. Under the new law no permit (which will include a compliance schedule) will be approved unless a credible source of financing exists; technical feasibility will no longer be enough.

Fourth, in the past regulations were often introduced with little more than hope that enterprises would comply. Of particular note is the current high percentage of enterprises operating without permits. To avoid the same result under the new framework environmental law, a transition scheme is included to avoid regulatory bottlenecks as thousands of enterprises file permit applications. This phase-in period will also ensure that enterprises obtain permits when they are required to do so and that each application will receive adequate analysis.

Fifth, in the past the public neither was consulted nor received access to information about environmental hazards. Indeed, environmental information was considered a state secret. Without public backing, though, the EPAs lost an important source of input and potential support for environmental protection. In the future, public participation and free access to environmental information will become the legal presumption, and indeed EIA procedures will rely heavily on public input.

Key to getting enterprises to be more active in addressing environmental problems will be the use and enforcement of compliance schedules. Under the new law, compliance schedules will have two sections, one for ongoing pollution which must be brought into compliance within five years, and one for remediation of past damages that will have incident-specific timetables. Moreover, the law explicitly requires that a facility be closed when enterprises refuse to take steps mandated in compliance schedules.

The way forward for an expanded use of economic instruments such as pollution charges is very murky. To date, the political fear has been that environmental charges will be viewed as additional taxes. The future use of air charges will be based on the framework environmental law passed in 1995, which is based explicitly on the polluter pays principle. However, with the exception of fees for permitting, there is little explicit in the new law as to how the principle should be implemented or by whom. The hope is that in the near future pollution charges can be introduced. Key to the acceptance of such charges by the public – and enterprises in particular – will be to use revenues as a source of environmental investment financing. If perceived as simply another general revenue device, public opposition will be too great and they will not be introduced.

Turning to the water sector, the draft water law fortunately should address a number of the problems raised earlier in this chapter. First, it includes the beneficiary pays principle so that in the future water prices will be based on cost-recovery requirements. Second, it creates river basin commissions (RBCs) made up of local users, providers and regulators. RBCs will act as mini-parliaments, setting prices and prioritizing and approving investment plans. The national water utility, AR, will act as a technical advisor to the RBCs. The hope is that by giving control of pricing to local bodies, the effectiveness of water pollution and abstraction charges will increase, payment arrears will decline and water prices will rise to play a larger role in funding distribution systems. A new national water and district heating metering law is under development which will go further than the existing law by requiring the installation of bulk meters between municipal lines and apartment blocks.

There are a number of additional challenges which apply to both the air and water sectors. First, the new permitting requirements, including the need for EIAs and financially feasible compliance schedules, will put severe demands on the environmental authorities. The legislation does authorize fees for permit evaluation services, but evaluations take time and the demand for permits will be cyclical with a boom starting once the regulations are promulgated. Second, existing public finance law makes it difficult for budgetary institutions to maintain 'extrabudgetary' funds. This will make it very hard to be sure that the new permitting system is adequately financed.

Third, a major barrier to creating an effective system is a lack of monitoring equipment, adequate laboratories and sufficiently trained staff (the best having left because of the erosion of public sector salaries and those remaining being overworked). This point is particularly true with respect to air pollution. Finally, fear of decentralization has made it hard to garner interest in more flexible permitting mechanisms (such as tradable permits) that could encourage lower-cost compliance, but which are essentially regional instruments.

Nevertheless, in spite of the many problems mentioned above, Romania has embarked on a radical course of environmental reform. Environmental policy reform is an area in which public interest is growing. The carrot of accession into the European Union has created a very encouraging new policy focus and impetus for action. The experiences of other countries suggest that such reforms take time to gather momentum. Romania has started down the right road, and a well-run permitting system – supported by a complementary set of pollution and abstraction charges – is the *key* to success.

REFERENCES

DeShazo, G., D. Cotiusca and C. Zinnes (1996), 'The Willingness to Pay for Urban Services in Romania', Harvard Institute for International Development, International Environment Program, Working Paper.

Government of Romania (1996), *Romanian Framework Environmental Law,* Monitorul Official No. 304, 30 December 1995, Part I, pp. 1–15, Government of Romania.

(ICIM) (1995), *Questionnaire on the Control of Industrial Discharges to the Environmental in the Danube River Basin.*

Manea, G. and C. Zinnes (1994), 'An Overview of the Impact of the Most Polluted Sectors in Romania', Harvard Institute for International Development, International Environment Program, Working Paper.

(MAPPM) (1994), *National Environmental Protection Legislation and Regulations,* Environmental Documentation and Information Office.

(MAPPM) (1995), *Report regarding the National Action Program for Environmental Protection (Synthesis),* Ministry of Waters, Forests and Environmental Protection.

National Commission for Statistics (1996), *Romanian Statistical Yearbook, 1995,* National Commission for Statistics.

Regional Environment Center (1994), *Use of Economic Instruments in Environmental Policy in Central and Eastern Europe,* Budapest: Regional Environmental Center, pp. 122–36.

World Bank (1992), *Romania Environmental Strategy Paper,* Report No. 106113–ROMANIA, Washington, DC: World Bank.

Zinnes, C. (1996), 'The Road to Creating an Integrated Charge and Permitting System in Romania', Harvard Institute for International Development, International Environment Program, Working Paper.

13. Implementing Pollution Permit and Charge Systems in Transition Economies: A Possible Blueprint

Randall Bluffstone and Bruce A. Larson

1 INTRODUCTION

Since seminal articles were published in the 1960s and 1970s (for example, Dales, 1968; Baumol and Oates, 1971; and Montgomery, 1972), there has been substantial analysis of economic instruments, and the call for their adoption and implementation has become well-entrenched in the policy arenas of many countries. For example, a workshop on economic instruments for sustainable development held in 1995 under the auspices of the United Nations Commission for Sustainable Development praised the use of these instruments. Recommendation 2.3.5 from this workshop suggested that the UN and international organizations should 'promote new applications for existing instruments and the development of new and innovative instruments for sustainable development' (Moldan (ed.), 1995, p. 3). Similar calls can be heard from international organizations such as the World Bank and the OECD, as well as from regional bodies (World Bank, 1992; Opschoor and Vos, 1989; and Nordic Council of Ministers, 1994).

In the forward to Moldan (1995), however, an important clarifying point is made: 'Everybody knows that command-and-control instruments are not the best solution and that economic instruments – taxes, charges, incentives, market based ones and others – are inherently better. However, in no country are economic instruments sufficiently being used at present' (p. 3). This may be somewhat of an overstatement, with more than 200 instances of various economic instruments being used in OECD countries, of which more than 100 of these are some type of charge (Potier, 1995), but the general point is well taken that instruments which look best on paper – direct charges on emissions and transferable emission rights – have not been widely implemented. As already mentioned in the introduction, in OECD countries in 1995 there were five cases where emissions or effluents were measured or estimated and, on the basis of those calculations, charges were assessed.

Based on the experience discussed and analysed in the various country chapters, the main purpose of this final chapter is to outline key lessons – somewhat of a blueprint – for the design and implementation of pollution charge systems. While this 'blueprint' is drawn from transition economies, and therefore many of the lessons are most relevant for such conditions, it should be remembered that a large portion of the world can reasonably be classified as undergoing some sort of transition to more market-oriented economies. Thus, to a very large degree the experience here is relevant not only for countries of Central and Eastern Europe and the former Soviet Union, but also for a substantial portion of the rest of the world.

As the various country-specific chapters in this book show, the experience from the period 1991–95 has been sufficient to allow at least a preliminary assessment of the lessons from the widespread use of pollution charges. Much has been learned about how to structure such policy instruments under less than ideal conditions in which pollution charges are largely 'add-ons' to existing regulatory environments based on performance standards. In several countries substantial revisions have already taken place. Indeed, how to integrate permit and charge systems, given the many obstacles, is perhaps the main lesson to be drawn from the country chapters.

The fact that these charges were incorporated into existing permit and standard systems should not be too surprising because that is how charges have been incorporated into most regulatory structures. As Hahn (1989) concluded in a well-known review of the use of economic instruments, 'virtually none of the systems . . . exhibits the purity of the instruments which are the subject of theoretical inquiry' (p. 97). Hahn also notes that combinations of effluent charges and permits are used in France, Germany and the Netherlands, and that in the United States: 'In all cases charges are added to the existing regulatory system which relies heavily on permits and standards' (p. 106). Indeed, his conclusion, written before the Berlin Wall came down, is almost tailor-made for this book:

> An examination of charge and marketable permits schemes reveals that they are rarely, if ever, introduced in their textbook form. Virtually all environmental regulatory systems using charges and marketable permits rely on the existing permitting systems. This result should not be terribly surprising. Most of these approaches were not implemented from scratch; rather, they were grafted onto regulatory systems in which permits and standards play a dominant role. (Hahn, 1989, p. 107)

2   WHAT CAN THE REST OF THE WORLD LEARN FROM THE TRANSITION COUNTRY EXPERIENCE?

To close this book, we attempt to draw out the lessons from the country chapters and provide our set of recommendations for designing the key features of a pollution control policy that includes a combination of pollution permits and charges. There is no attempt here to 'prove', either theoretically or empirically, that these are the best solutions to the many problems that exist. It is our best attempt at distilling an emerging set of practical experience, guided by basic economic logic, combined with a healthy respect for the difficult job facing environmental authorities in transition economies. It will perhaps come as no surprise that a common feature of these recommendations is that they are designed to economize on transaction costs associated with developing, implementing, and enforcing pollution control policy.

a.   Simplify systems and focus attention on the most important pollutants
To target pollution charge systems more precisely and to make the most of scarce human and financial resources, pollution charge systems should focus primarily on a relatively small group of 'main' pollutants. These pollutants about which systems are most concerned should be – and in countries that have made such simplifications usually are – common, important pollutants that are largest in volume, and are ones for which at least end-of-pipe control options are more-or-less known. If an objective of using charges is to create incentives for cost-effective pollution abatement of these pollutants, they should also be less toxic pollutants emitted at levels that do not generate large environmental damages. For such pollutants it will therefore be reasonable to offer the flexibility necessary for improving cost effectiveness.

Pollution charge systems in the countries included in this book not only cover the most important pollutants, but indeed tend to be comprehensive or very nearly so. In most countries well over 100 pollutants are subject either to charges or penalties or both. To deal with the complexity of choosing such a large number of rates, many countries have defined three or four toxicity groups in which all pollutants in a group are charged the same rate and emissions of more-toxic pollutants are always charged higher rates than less-toxic ones. If political forces dictate that a variety of pollutants must be charged, such methods can transform very unwieldy systems into more manageable ones with relatively few rates. This simplicity, of course, comes at a cost. Because each category may have 25 or more pollutants, charges cannot be expected to achieve particular environmental results for more than a few of these pollutants. The purpose of charging toxicity groups at all is therefore mainly to raise revenues, and as will be discussed in (d) below, to enforce emission limits.

b.   *Choose national/regional policy objectives in terms of aggregate emissions levels or aggregate emission reductions linked to ambient environmental quality goals*

When looking at instruments for environmental policy, it is important to distinguish between the environmental objectives and the means to achieve those objectives. However, except for a small set of pollutants, usually related to international conventions, few countries explicitly state environmental policy objectives in terms of simple and verifiable goals. While implicitly all current ambient standards could be taken directly as policy objectives, the network of monitoring stations is, and for the foreseeable future will remain, inadequate for monitoring many common pollutants, and will be non-existent for more exotic and toxic pollutants. It is therefore almost impossible to determine if current policies or proposed changes in policies will be adequate to achieve national goals.

Policy makers should therefore focus on setting policy goals in terms of emission reductions. Given such goals, environmental policy can then concentrate on the question of finding cost-effective solutions to meet those objectives. A point made by Hahn and Stavins (1991, p. 29) with respect to the US is particularly relevant for this discussion,

> Separation of goals and standards from the means of achieving those goals and standards holds symbolic importance. Implicit within the current round of incentive-based recommendations is the notion of using the conventional deliberative process to establish goals and standards, while achieving those standards by the least-cost means.

c.   *Set annual performance standards codified in permits by pollutant and by facility, not by individual source (for example, by stack)*

At minimum, a bubble approach should be used for each facility where aggregate emissions are regulated, rather than emissions by source within the facility. This approach reduces administrative costs for both enterprises and local environmental authorities, and at worst has a neutral environmental impact. It may, however, allow firms to reduce their compliance costs through within-facility trades.

It is possible that in some circumstances an aggregate facility bubble would not be appropriate, in which case some subset of sources could be combined into one bubble. For example, at a major cement plant in

Estonia there are multiple 'high' stacks and multiple 'low' stacks. The high stacks have long-range impacts, while the low stacks have very local effects on workers and the city in which the plant is located. In such cases two limits would be reasonable, but in general the presumption should be that one limit is adequate unless otherwise obvious. It should also be noted that, to some degree, this approach is already implicitly followed in several countries by regional environmental authorities who just do not have time to worry about each source.

d.   Choose a core set of priority air and water pollutants and a two-tiered charge structure that is linked to facility performance limits
In all countries considered here, charge systems are integrated into exist-ing permit structures and they are also typically called upon to do at least two jobs at once. Two-tiered rate structures, with base rates for pollution up to limits and higher penalty rates for emissions above limits, not only fit well with existing permit systems, but they can make practical sense when there is a need to deal with many different types of pollutants simultaneously. For example, with regard to so-called main pollutants, the goal of charges might be to provide textbook-type incentives to reduce emissions. Under such circumstances, base rates are most important, with penalties that are small multiples of base rates and limits playing secon-dary roles. Charge systems also, however, can contribute to controlling pollutants that are believed to have high marginal damages (for example, ones that are inherently very toxic). In such cases penalty rates are most important, because the goal is not to allow flexibility, but to keep emis-sions below dangerous levels as defined by limits. Penalty rates therefore serve to enforce limits and base rates only generate revenues.[1]

Even if we consider only main pollutants, though, a two-tiered charge structure can be a practical structure when there are uncertainties about abatement costs and environmental damages. When charges are sup-posed to help achieve quantifiable objectives as suggested in (b), using only uniform charges puts a weighty responsibility on the analyst to set rates correctly even when key information is likely to be missing. Bound-ing what are believed to be 'correct' uniform charge rates with base and penalty rates can therefore provide useful insurance against mis-esti-

---

1   A potential disadvantage of charge systems is the difficulty of adapting a national charge to regional differences in damages. Although in principle it would be possible to set special charge rates for different regions and in the extreme for every polluter, in practice countries have employed regional multipliers when they want to discourage emissions by different amounts in different areas. Examples where this tool is used include Russia, Slovakia, Lithuania, Estonia and, though not in this book, Kazakhstan (Vasiliev, 1995). Another possibility is that limits can vary across regions so that the *i*-th unit of pollution in one location triggers the base rate, and in another location results in the penalty rate.

mates. If trading in emissions limits is allowed, such systems can even potentially outperform either pure charges or pure tradable permits in terms of cost effectiveness (Roberts and Spence, 1976).

Thus, a reasonable answer seems to be emerging regarding how to develop sensible, but still rather comprehensive charge systems. For main pollutants, base rates are where emphasis is placed in terms of calculation effort and penalty rates should be relatively small multiples of base rates. It is with regard to these pollutants that flexibility in pollution abatement is encouraged and the gains from economic instruments are expected. For non-priority pollutants, and particularly for more hazardous emissions, the concept is quite different. Base rates are set to raise revenues and are not really meant to influence behaviour *per se*. The main emphasis is placed on calibration of penalty rates that provide sufficient incentives for polluters to keep emissions below limits. Because toxicities vary, however, penalty rates should not be the same multiples of base rates for all pollutants.[2]

e.   *Where possible, use abatement costs to guide the choice of charges*
      and the jump between base and penalty rates

As was discussed in Chapter 1, the desirability of pollution charges as an economic instrument is predicated on the idea that polluters know abatement costs, but as several of the country chapters suggest, it is far from clear that this is the case. If policy makers were omniscient and knew themselves what abatement cost schedules looked like, this would not matter; polluters would simply be induced to engage in an optimal search for information. Unfortunately, analysts working in transition economies must rely on these same polluters (who do not know their least-cost abatement options) for information on which to based charge rates.[3]

Focusing policy objectives on emission reductions, as recommended in (b) above, greatly simplifies the question of how to set charge rates, because marginal damages that differ by pollutant and to some extent by region are dealt with by varying emissions goals and facility limits. Indeed, a true cost–benefit calculation is not needed, because in principle the charge should be set where marginal abatement costs equal the charge at the right level of emissions. The recurring problem, of course, is that it is impossible to find even one point on average abatement cost curves for many of these

2   This is the approach which has been taken in Estonia, where penalty-rate multiples of base rates vary according to the toxicities of pollutants emitted.

3   As was recognized by Baumol and Oates (1971), the requirement that regulators know abatement costs, which is private information, in order to set charge rates properly is paradoxical at best. Perhaps even more fundamental, the answer to the question 'what are marginal abatement costs?' is a moving target that must be estimated today, but will continue to change over time as technologies change.

pollutants, let alone complete marginal abatement cost schedules. At the same time, especially for remaining state-owned enterprises and their private sector spin-offs acquired by past managers, employees and government officials, there may be no real need to understand least-cost pollution-control options because they are not cost minimizers.

It should also be recognized that the literature is extremely thin on the question of abatement costs. There are surprisingly few studies of firm-level abatement costs where more than a few pollutants are covered and heterogeneous industries are considered. Perhaps one of the best data sets is from the annual US Department of Commerce Pollution Abatement Cost and Expenditures Survey, but this allows estimates only for a few specific pollutants and the rest are aggregated into two toxicity-based categories (US Department of Commerce, 1994). Hartman et al. (1994), however, show that some interesting analysis is still possible, though their analysis focused on the years 1979–85 when somewhat more detailed information was available.

Within the countries in transition, some estimates of abatement costs (that is, single points on abatement cost schedules) exist for Poland, but these are generally regional or sectoral studies and are largely restricted to end-of-pipe methods.[4] Transferring such estimates is questionable because of substantial cross-country differences, but particularly when estimates from other countries are lacking one perhaps does not want to bet too heavily on the accuracy of abatement cost estimates. Charging more-toxic pollutants at higher rates than less-toxic ones, as discussed in (a), and using two-tiered structures as suggested in (d) may therefore be considered reasonable.

f.   Develop cost-effective and non-adversarial approaches to implementation and enforcement

Throughout this book, cost effectiveness is discussed primarily with respect to the structures of pollution charge systems, but perhaps not enough has been said about cost-effective implementation and enforcement of these systems. There are two related issues here: coverage and enforcement. The first issue – the coverage issue – deals with which polluters should be included in the system of permits and charges. Most countries have said that in principle 'all' polluters should be permitted and charged for specific pollutants if they pollute above some minimum level. The question of appropriate minimum levels is therefore an important policy question that, perhaps, has not been adequately addressed.

---

4   For example, see Adamson et al. (1995); Berbeka (undated); Zylicz (1994); Broniewicz et al. (1994).

In Latvia and Estonia, for example, a very small number of large polluters account for the vast majority of stationary source emissions. The
same is true in parts of Russia. In such situations, there may be little
benefit and probably net social costs of fully including small polluters in
permit and charge systems. Some streamlined permitting process based
on simple self-reports for information purposes could be a middle ground
between ignoring small polluters altogether and fully including them in
charge and permit systems. Such polluters should probably also be exempt from charges if the charge revenues do not justify the administrative
costs associated with collecting and recording transactions.[5] Such regulatory tiering, which is common in some parts of the world, would be a
useful way to focus attention and regulators' efforts clearly on a core set
of important polluters.

Distinct from 'who' should have permits and pay charges, the second
issue – enforcement – pertains to how environmental authorities should
encourage compliance with the system of permits and charges. In simple
models it is simply assumed that it is possible to monitor and enforce
systems of permits or taxes, but in reality the existence and level of
charges, penalties and fines associated with violations of permit conditions do affect incentives for compliance and evasion.[6]

It is clear from the country chapters that constraints associated with
inspection activities are tight in terms of both money and trained person-
nel. One way to economize is to reduce the coverage of systems, thereby
reducing the number of enterprises that need to be monitored and/or
inspected by environmental authorities. Given the number of polluters in
the system, a cost-effective enforcement strategy can then be considered.
One approach that seems to be used is to focus enforcement attention on
the largest polluters in the first place by organizing more frequent inspections, including unannounced visits. In many cases, however, the 'biggest'
polluters comply with permit and charge requirements, while the rest of
the pack has more varied compliance. An alternative discussed in Harrington (1988) and Ubelis et al. (1996) is to link reporting requirements
of enterprises and the frequency of inspections to enterprises' compliance
histories. The 'good' group would have less frequent inspections, while the
'bad' group would have more frequent inspections. Over time enterprises
move between the two groups.

5   As discussed in, for example, Bluffstone and Varneckienė (1996) and Ensmann et al.
(1995), administrative costs are not closely linked to the presence of a charge system *per se,* but
they are related to the coverage of the system and especially the associated system of permits.
6   Note that compliance in this context implies truthfully reporting emission levels and
paying appropriate charges on time, but not necessarily polluting within emission limits in
permits.

Regardless of the enforcement strategy, given the relatively weak positions of environmental regulators in most countries, except for a few high-profile cases where plants have been closed for extreme reasons, it is desirable for environmental authorities not to develop adversarial relationships with polluters. The authors have witnessed first-hand the tensions that can develop during inspections. Such confrontations yield few benefits and create a lingering unease that is not conducive to generating industry acceptance of pollution-control policies. At the same time, announcing inspections too far in advance probably allows too much room for cleanup before inspections. A reasonable middle ground, as suggested in Ubelis et al. (1996) involves a combination of regularly scheduled visits and 'spot' inspections held after giving, perhaps, one day's notice.

g. *Use penalty charge rates to define levels of liability for accidental discharges and for deliberate evasion*

In transition economies it is almost impossible for environmental authorities, except in the most extreme situations, to use non-administrative approaches, such as civil and criminal courts, to enforce environmental policies. Instead of using such procedures, the standard method for dealing with environmental accidents has been to use complex and perhaps poorly-designed systems of 'damage methodics' to set fines that responsible parties must pay. These methodics were created under central planning and therefore were not designed for use under market economy conditions. They are also probably not appropriate for use in the contemporary economic situation.[7]

At the same time, however, environmental authorities do not have the skills and/or resources to hire skilled specialists to conduct market-oriented damage assessments for each environmental problem, and it is not clear that this is even a good idea. In several countries the system of pollution charges has simplified the system of assessing liability by defining such accidents or evasions as 'above-limit' emissions, with cleanup costs treated as a separate issue. This is a reasonable use of the charge system and provides some internal consistency within the environmental policy framework.

h. *Integrate the system of pollution charges into the general system of income/profits taxation.*

As part of a possible future 'greening' of tax policies, it is desirable to integrate the system of pollution charges into the general tax system. This has two implications. First, at least the process of paying pollution charges should be incorporated into the general system of income/profit taxes. There is no need to set up parallel collection systems, and it seems

---

7 It should be noted that in some countries these methodics are still used.

reasonable that tax authorities can be responsible for collecting charges. Because these charges are deductible expenses for tax purposes, the two types of payments are also conceptually linked.

To be sure, tax authorities will not want the work, especially if pollution charges are not considered to be part of general revenues. Tax authorities will also not want to be responsible for determining if reported emissions and calculated charges are correct. A happy middle ground could be for environmental authorities to be wholly responsible either for determining charge payments or for providing 'spot' checks of tax declarations to see if enterprises are calculating charges correctly. They could also assist enterprises by checking their charge calculations before the forms and payments are sent to tax authorities.

A second point is that all pollution charges, both for within- and above-limit emissions, should be deductible for tax purposes. In most countries, payments for within-limit emissions are deductible, but charges on above-limit emissions are not. There is no clear logic for this approach, although the marginal incentives of the penalty charge rates are increased by non-deductibility conditions.[8] Again, ministries of finance will not like such provisions when pollution charge revenues are not directed to general revenues.[9]

i.   *Charge levels must be clearly indexed for inflation, and such indexing must automatically occur each time period (for example, year, quarter)*
Experience on this point is clear. Serious inflation, a reality in transition economies especially during early stages, erodes any incentive effects of charges as well as the real value of revenues collected. At the same time, the periodic large increases in charges rates (after a few years, for example) that have occurred to try to catch up with inflation, do not provide smooth signals to enterprises about real charge rates. Thus, specifying known periods and methods of indexing (for example, quarterly or annually-based on changes in the producer price index) is absolutely necessary.

j.   *Creating some form of pollution charge waiver and 'environmental fund' is probably a political necessity in response to distributional concerns associated with charges*[10]
While most environmental economic analysis related to pollution charges focuses on determining the 'right' charge rate and its effects, it is well

8   More discussion of this point is included in the chapter on Estonia.
9   It should not be forgotten that depreciation allowances in income tax systems and VAT may have more impacts on the final cost of pollution-control investments than pollution charge payments.
10   It is perhaps important to make a clear distinction between the general term *waiver* and the more specific terms *allowance* and *credit*. According to Webster's Dictionary (1965),

recognized in transition economies that 'where the revenues go' is of central importance for the political acceptability of charge systems. Although cost effectiveness is key, decisions regarding the distribution of costs and benefits from using economic instruments probably determine whether such instruments are ever implemented. It must be recalled in particular that pollution charges only save money in the sense that they reduce the aggregate costs of pollution control, and in most cases individual polluters pay more with pollution charges than with performance or technical standards. The use of charge revenues, therefore, becomes central to the discussion in finance-starved economies.

Industrialists must be convinced, particularly, that the benefits of increased flexibility outweigh the costs of increased payments to governments. The only way to do this is, of course, to channel most or all of the collected charges back to the polluting community. Such a distribution scheme is perhaps reasonable, because even with a relatively broad definition of the term environmental investment there is probably little or no finance available outside of individual polluting firms themselves. Capital markets are simply not yet sufficiently developed to provide this type of financing. Terms tend to be short, with high real interest rates and probably high levels of risk as well. Nascent banking sectors do not have the experience to analyse such projects and potential borrowers do not know how to package them. For these and other reasons, little finance is provided for environmental investments.

Two related approaches, the use of pollution charge waivers for environmental expenses and the creation of environmental funds for centralized investment decision making, have become common responses to distributional concerns associated with pollution charges. These methods are probably necessary evils. A system of pollution charge waivers, specifically credits based on very simple and observable criteria, is perhaps the preferred approach. The goal should be to try to mimic standard investment tax credits, such as investment depreciation allowances in standard tax systems. This approach keeps money in the hands of polluters and does not involve having them pay only to receive it back again

---

a *waiver* involves the 'act of intentionally relinquishing or abandoning a known right, claim, or privilege or an instrument evidencing such an act'. One type of waiver involves an *allowance,* which is a 'sum granted as a reimbursement'. For example, a pollution charge waiver is an allowance when the polluter has to pay pollution charges first, and then try to recover that amount during its investment planning process. Another type of waiver is a *credit,* which is a deduction from an amount otherwise due. With a credit approach to a pollution charge waiver, the enterprise is granted the waiver before any payments are made and, as a result, does not have to try to recover its money from some governmental institution. Needless to say, enterprises probably favour credit approaches.

through some distribution mechanism. The approach therefore economizes on transaction costs. To be sure, systems of pollution charge credits do not ensure cost effectiveness, but they at least somewhat make up for this disadvantage by explicitly recognizing the perhaps inevitable inefficiencies in centralized environmental investment funds found in the region.

When pollution charge payments are significant, waivers may provide incentives for both emission reductions and – at least within one firm – incentives for using scarce funds in the best way. Often, however, rather than establishing clear rules regarding when waivers are allowed, substantial discretion has been built into systems. The authors have direct experience with one country where the regulators attempted to draft a clear regulation on such waivers, but gave up in frustration when it became too difficult to determine identifiable criteria – that were still acceptable to environmental authorities – under which waivers would be granted. Keeping the decision process blurry made it easy for the regulators.

As discussed in the country chapters, in Poland and Hungary regulatory structures allow penalties to be waived if enterprises show evidence that they are making investments to overcome noncompliance. In practice, however, this implicitly occurs in several countries, because environmental authorities do not want to place large financial burdens on enterprises; granting such waivers becomes particularly likely when regulators know they may not get the money. Afterwards they are often simply considered to be part of the overall system of investment tax allowances.

Various forms of 'environmental funds' financed by pollution charges, both persist and grow in many countries in the region.[11] These environmental funds have their origins during central planning times and often are governed by 'old' thinking that centralized institutions are the best way to make environmental investments. Environmental authorities in the past directed systems of environmental investments, and the notion of contemporary environmental funds at least partially derives from such earlier systems. It is also true, however, that multiple purposes and forces are behind the development of environmental funds, and have determined their uses and results in various countries. To some degree, these funds may be used to overcome the capital market constraints that oth-

11   Historically, at least in the former USSR, the real purpose for implementing pollution charges during the period of central planning was to acquire resources to pay for local/regional pollution-control investments. As direct budget allocations to make such investments began to disintegrate in the 1980s, pollution charges were viewed as an alternative mechanism to acquire investment resources. In fact, as discussed in the chapter focusing on Russia, charge levels were related to the magnitude of desired state-directed pollution-control investments.

erwise stall environmental investments. Some examination of the experience with these funds has taken place in recent years, and it is perhaps fair to say that a set of good operating practices has been developed and widely discussed in the region (for example, Lehoczki and Peszko, 1994, 1995; Lovei, 1994; CowiConsult, 1995).

Many country chapters have discussed the operation of environmental funds. Without going into excessive detail, it perhaps can be said that operation of these funds is certainly at a qualitatively lower level than one would hope. At the national level at least, they tend to be divisions of ministries of environment and, only partly for this reason, are typically seen as institutions that further the goals of environment ministries rather than ones designed to overcome capital market failures. Close examination of these funds often reveals a surprising lack of clarity regarding operating practices, and an amazing lack of evaluation of the environmental and economic effects of the investments that are made.

In cases where pollution charges are earmarked to environmental fund systems, charge revenues are often derived mainly from private sector or public enterprises subject to hard-budget constraints. On the distribution side, however, one East European lawyer perhaps best stated the general problem of politicization of environmental funds and the tendency to disburse grants rather than loans, when he noted that 'everyone likes to give gifts. It makes us feel good'. The focus on giving gifts rather than overcoming capital market failures has meant that project appraisal processes are generally not transparent and the reasons for distributing funds are not based on objective environmental and economic criteria. One manifestation of this phenomenon is that distributions tend to be at the very least tilted towards municipal and other state enterprises (Farrow and Bluffstone, 1994), and in some cases fund revenues can even be used for incentives for ministry of environment inspectors or to support various research programme.

Municipalities often demand a piece of the pie explicitly and in several countries so-called municipal or regional funds have been created. In countries where total charge revenues are small, however, the amounts which municipal funds receive may be only fifty or sixty thousand dollars, and in some cases only a few thousand dollars per year. It is therefore difficult for them to do something substantial in the environment sector and expenditures tend to be 'defined' as being for environmental projects, but in practice these resources are treated like general budget revenues.[12] Problems on the distribution side therefore tend to be even larger in municipal

---

12 For example, when fund revenues are used to buy computers for processing pollution permit and charge data, or for public area cleanup and landscaping.

funds than in national funds and often the use of these resources at the local level is not clearly understood, monitored or controlled.

Over time, however, both environmental funds and pollution charge waivers may create identifiable projects for which environmental outcomes are known. This can economize on monitoring and enforcement effort, because these instruments offer incentives for enterprises to reveal private information about their activities. They also provide at least one set of estimates of abatement costs that can be used to set appropriate charge rates (Zylicz, 1994), although this presumes that definitions of environmental investments are broad enough to encompass both end-of-pipe investments and process changes. Poland has also attempted to partly address its collection problem by linking payment of charges with access to the resources of its national environmental fund. Without having paid all charges and penalties, enterprises cannot receive loans from the national fund. Latvia has taken this a step further by requiring that an enterprise has no payment arrears of any kind with the government.

3 THE FUTURE

The desired accession to the European Union for most countries will overlay all other efforts in the foreseeable future. A major area where this trend will be seen is in the area of ambient and enterprise-level standards. As was mentioned in several chapters, it is often the case that country ambient standards are stricter than standards in the European Union or its member countries. Because limits in permits are often linked to ambient standards, and because limits play a large role in determining charge payments, relaxing overly strict standards will reduce undue regulatory burdens on enterprises and increase the credibility of existing systems. To the extent that European Union enterprise-level standards based on best-available control technologies are adopted, however, this will represent a pernicious development by forcing overly expensive technologies on transition economies. It will also cause these countries to pass up low-cost/high-benefit projects that would otherwise be attractive.

Pollution charge systems combined with performance standards are fundamentally monitoring and enforcement-intensive instruments best suited to pollutants where it is desirable to offer flexibility regarding abatement levels. As revenue contributions of minor pollutants shrink relative to criteria pollutants, however, and as it is increasingly recognized that the transaction costs associated with pollution charge systems are not trivial, it is likely that pollution charge systems will become less comprehensive and will include fewer pollutants and fewer polluters. As part of this trend, it is perhaps reasonable to expect an increased reliance on instruments which less intensively use scarce monitoring and enforcement capability, but which may also be cruder tools. Particularly given the extensive experience in Western Europe with product and polluting input charges, such instruments are likely to complement, or in some cases displace, pollution charges.

It is clear that the issue of distribution of revenues from pollution charges will continue to loom large, particularly as system structures are improved and are able to generate larger revenue streams. In countries where charges now go to national or local government budgets, battles are on the horizon with those who would prefer to channel a greater portion to enterprises. In countries where decisions have been made to allocate the majority or all of charge revenues to environmental fund systems, the major issue will be the improved performance of those funds and the development or refinement of procedures for co-financing large activities. As economies continue to change and private sectors expand, it will be particularly important to orient fund systems to the needs of enterprises and to reduce existing biases towards funding municipal environmental projects.

Controlling transaction costs associated with implementation of environmental policies in transition economies is a key issue given the limited budgets, staff and capacities of people available to carry out a wide variety of duties. As was emphasized in practically every chapter in this book, monitoring and enforcement are considered deficient throughout the region. Adding charges to the policy framework almost certainly increases incentives for evasion and therefore drives up these enforcement costs. Indeed, in Romania, Slovakia and Russia charge collections are less than half of charge liabilities, and evidence from both Romania and Poland suggests that collection rates decline as charge rates increase. In Lithuania, one major water utility reported that 1994 pollution fine collections were only 15 per cent of what was assessed.

While most environmental authorities think that their pollution charges systems increase incentives for investment in pollution control, very little evidence exists that charge systems actually do provide such incentives. Investment decisions in these regions take place in highly uncertain market and regulatory environments, and under such conditions the relationships between current charge levels, future charge levels and investments in pollution control are not simple.[13] Almost nothing has

13   In addition to the brief presentation in the appendix to Chapter 1 of this book, Magat, 1978; McCain (1978); Pindyck (1991) and Larson and Frisvold (1996) also discuss this problem.

been said about these dynamic issues in this book and it is a major area for future theoretical and applied work.

REFERENCES

Adamson, S., R. Bates, R. Laslett and A. Pototschnig (1995), 'Economic Instruments and Environmental Policy in Krakow: An Evaluation', Paper Presented at the Sixth Annual Conference of the European Association of Environmental and Resource Economists, Umea, Sweden, June.

Baumol, W. and W. Oates (1971), 'The Use of Standards and Prices for Protection of the Environment', *Swedish Journal of Economics,* March: 42–54.

Berbeka, K. (Undated), 'Assessment of the Direct Costs Caused by the Reduction of Nitrogen and Phosphorous Discharges into the Baltic Sea', Mimeo, Warsaw Ecological Economics Center, Warsaw University.

Bluffstone, R., and J. Varneckienë (1996), 'The Impact of Transactions Costs on the Effectiveness of Pollution Charges in Central and Eastern Europe', Mimeo.

Broniewicz, E., B. Poskrobko and T. Zylicz (1994), 'Internalizing Environmental Impacts of Industry in Poland: Preliminary Empirical Evidence', Working Paper, Warsaw Ecological Economics Center, Warsaw University, June.

CowiConsult (1995), 'Establishment of a New Environmental Revolving Fund', Prepared for the Ministry of Environmental Protection and the European Commission, September.

Dales, J.H. (1968), 'Land, Water and Ownership', *Canadian Journal of Economics,* November: 791–804.

Dudek, D., Z. Kulczynski and T. Zylicz (1992): 'Implementing Tradable Rights in Poland: A Case Study of Chorzow', in *Proceedings of the Third Annual Conference of the European Association of Environmental and Resource Economists,* Vol. 2, pp. 58–75.

Ensmann, E., L. Gornaja and B.A. Larson (1995), 'The Pollution Policy "Cocktail" in Estonia: Economic Incentives and Problems for Implementation', Environmental Discussion Paper No. 5, Harvard Institute for International Development, January.

Farrow, R.S and R. Bluffstone (1994), 'Implementable Options for Cost Effective Air Pollution Reductions in Northern Bohemia', Report prepared for the Ministry of Environment of the Czech Republic and the Harvard Institute for International Development.

Fournier, B. (1995), *Environmental Taxes in OECD Countries,* Paris: OECD.

Hahn, R.W. and R.N. Stavins (1991), 'Incentive-Based Environmental Regulation: A New Era from an Old Idea', *Ecological Law Quarterly,* **18** (1): 1–42.

Hahn, R.W. (1989), 'Economic Prescriptions for Environmental Problems: How the Patient Followed the Doctor's Orders', *Journal of Economic Perspectives,* **3** (2), Spring: 95–114.

Harrington, W. (1988), 'Enforcement Leverage when Penalties are Restricted', *Journal of Public Economics,* **37** (1): 29–53.

Hartman, R., M. Singh and D. Wheeler (1994), 'The Cost of Air Pollution Abatement', World Bank Policy Research Working Paper 1398, World Bank, December.

Larson, B.A., and G.B. Frisvold (1996), 'Uncertainty over Future Environmental Taxes: Impact on Current Investment in Resource Conservation', *Environmental and Resource Economics*, **8**: 461–471.

Lehoczki, Z., and G. Peszko (1994), 'Environmental Funds in the Transition to a Market Economy', Paper Prepared for the OECD Conference on Environmental Funds, St. Petersburg, Russia, October.

Lehoczki, Z. and G. Peszko (1995), *The St. Petersburg Guidelines on Environmental Funds in the Transition to a Market Economy*, Paris: OECD.

Lovei, M. (1994), 'Environmental Financing: The Experience of OECD Countries and Implications for Transition Economies', Paper Prepared for the OECD Conference on Environmental Funds, St. Petersburg, Russia, October.

Magat, W.A. (1978), 'Pollution Control and Technological Advance: A Dynamic Model for the Firm', *Journal of Environmental Economics and Management*, **5**: 1–25.

McCain, R.A. (1978), 'Endogenous Bias in Technical Progress and Environmental Policy', *American Economic Review*, **68**: 538–46.

Moldan, B. (ed.) (1995), *Economic Instruments for Sustainable Development*, Ministry of Environment of the Czech Republic.

Montgomery, W.D. (1972), 'Markets in Licenses and Efficient Pollution Control Programs', *Journal of Economic Theory*, December: 395–418.

Nordic Council of Ministers (1994), *The Use of Economic Instruments in Nordic Environmental Policy*, Copenhagen.

Opschoor, J.B. and H.B. Vos (1989), *Economic Instruments for Environmental Protection*, Paris: OECD.

Pindyck, R.S. (1991), 'Irreversibility, Uncertainty, and Investment', *Journal of Economic Literature*, **29**: 1110–48.

Potier, M. (1995), 'The Experience of OECD Countries in their Domestic Use of Economic Instruments for Environmental Management', in B. Moldan (ed.), *Economic Instruments for Sustainable Development*, Ministry of Environment of the Czech Republic, pp. 58–65.

Roberts, M.J. and M. Spence (1976), 'Effluent Charges and License Under Uncertainty', *Journal of Public Economics*, **5**: 193–208.

Ubelis, Arnolds, Valdis Seglins and A. Malik (1996), 'An Assessment of Selected Policies for Controlling Stationary and Point Source Pollution Emissions in Latvia', draft report for the Harvard Institute for International Development C4EP Project, Cambridge, MA, 9 July.

US Department of Commerce, Bureau of the Census (1994), *Pollution Abatement Costs and Expenditures, 1993*, US Bureau of the Census, December.

Vasilev, NA. (1995), 'Presentation at the Workshop on Implementation of Pollution Charge Systems in Transition Economies', organized by the Harvard Institute for International Development, Vilnius, Lithuania, September.

Webster's Seventh New Collegiate Dictionary (1965), Springfield, MA: G.&C. Merriam Company, Publishers.

World Bank (1992), *World Development Report*, Washington, DC: World Bank.

Zylicz, T (1994), 'A Survey of the Cost-Effectiveness of Investment Projects Co-Financed by the Polish National Fund for Environmental Protection', Working Paper, Warsaw Ecological Economics Center, Warsaw University, October.

Index